Electric and Hybrid Vehicles

T0253412

Electric and hybrid vehicles are now the present, not the future. This straightforward and highly illustrated full-colour textbook is endorsed by the Institute of the Motor Industry (IMI) and introduces the subject for further education and undergraduate students as well as technicians and workshop owners, with sections for drivers who are interested to know more.

This new edition contains extensively updated content, especially on batteries, charging and the high-voltage pathway and includes all new case studies and new images, photos and flow charts throughout. It covers the different types of electric vehicle, costs and emissions and the charging infrastructure before moving on to explain how hybrid and electric vehicles work. A chapter on electrical technology introduces learners to subjects such as batteries, control systems and charging, which are then covered in more detail within their own chapters. The book also covers the maintenance and repair procedures of these vehicles, including diagnostics, servicing, repair and first-responder information.

The book is particularly suitable for students studying towards IMI Level 1 Award in Hybrid Electric Vehicle Awareness, IMI Level 2 Award in Hybrid Electric Vehicle Operation and Maintenance, IMI Level 3 Award in Hybrid Electric Vehicle Repair and Replacement, IMI Level 4 Award in the Diagnosis, Testing and Repair of Electric/Hybrid Vehicles and Components, IMI accreditation, City & Guilds (C&G) and all other EV/hybrid courses.

Tom Denton is the leading UK automotive author with a teaching career spanning from lecturer to head of automotive engineering in a large college. His numerous bestselling textbooks have been endorsed by all leading professional organisations and are used by automotive students and technicians across the world.

Hayley Pells is Policy Manager at the Institute of the Motor Industry (IMI), an experienced MOT tester, and owner of award-winning Avia Sports Cars Ltd. She is a versatile automotive aftermarket writer, regularly published since 2007, an authoritative public speaker and consultant specialising in digital literacy, blended learning and electric vehicle technology.

Electric and Hybrid Vehicles

Third Edition

Tom Denton and Hayley Pells

Routledge
Taylor & Francis Group

LONDON AND NEW YORK

Contents

Contents

Preface

This third edition of *Electric and Hybrid Vehicles* is aimed at students who are taking a course in automotive technology or are interested in learning for themselves about this fascinating industry.

I am delighted to be partnering on this book with my friend and colleague Hayley Pells, a multi-award-winning owner of a highly respected service and repair business, Avia Autos, in Bridgend, South Wales, and the Institute of the Motor Industry's policy and public affairs lead.

Comments, suggestions and feedback are always welcome at my website: www.tomdenton.org. Several other books are available too:

▶ *Automobile Mechanical and Electrical Systems*
▶ *Automobile Electrical and Electronic Systems*
▶ *Automobile Advanced Fault Diagnosis*
▶ *Alternative Fuel Vehicles*
▶ *Automated Driving and Driver Assistance Systems*
▶ *Automotive Technician Training*

We never stop learning, so I hope you find automotive technology as interesting as we still do.

About the authors

Tom Denton has been researching and writing best-selling automotive textbooks for over 30 years. His published work is endorsed by all leading professional organisations and used by automotive students across the world. He has taught college students at all levels and always helped them achieve the best results. Tom was also a staff tutor and an associate lecturer for the Open University.

His postgraduate education in all aspects of technology and education, as well as many years of practical experience, has given him a broad base from which to approach and teach this interesting yet challenging subject.

As a fellow of the Institute of the Motor Industry and a member of the Institute of Road Transport Engineers and the Society of Automotive Engineers, he keeps in contact with the latest technologies and innovations across all aspects of this fascinating industry.

As well as publishing 45 textbooks, Tom has created amazing support materials and eLearning courses.

Hayley Pells expresses her gratitude for the opportunity to collaborate with Tom Denton, an esteemed technical author and experienced automotive engineer, who serves as a friend and inspiration. Throughout her time in the motor industry, which started operating an independent workshop, she discovered that the versatility of skills learned can be applied to both work and personal life. As a professional writer, she is published in many industry journals offering insight and information about the sector. She is respected for her efforts championing accessible education working with the upskilling of existing professionals, post-16 apprenticeships and numerous pre-16 careers and enterprise projects which showcase motor industry careers. She is a fellow of the Institute of the Motor Industry, holding their Patron's Award recognising her work campaigning for gender equality.

About the authors

www.tomdenton.org

There are extensive updates in this third edition. Thank you to friends and colleagues who sent feedback or made suggestions.

On our website, you will also find lots of free online resources to help with your studies. These resources work with the book and are ideal for self-study or for teachers helping others to learn. The resources include over **20 videos** to support your EV learning journey.

Acknowledgements

Over the years, many people have helped in the production of my books. I am, therefore, very grateful to the following companies who provided information and/or permission to reproduce photographs and/or diagrams:

AA
AC Delco
ACEA
Alpine Audio Systems
Audi
Autologic Data Systems
BMW UK
Bosch
Brembo Brakes
C&K Components
Citroën UK
Clarion Car Audio
Continental
CU-ICAR
Dana
Delphi Media
Denso
Department for Transport
Draper Tools
Eberspaecher
First Sensor AG
Fluke Instruments UK
Flybrid Systems
Ford Motor Company
FreeScale Electronics
Garage Hive
General Motors
GenRad
Google (Waymo)
haloIPT (Qualcomm)

Hella
HEVRA
HEVT
Honda
Hyundai
Institute of the Motor Industry
    (IMI)
Jaguar Cars
Kavlico
Ledder
Loctite
Lucas UK
LucasVarity
Mahle
Maserati
MATLAB/Simulink
Mazda
McLaren Electronic Systems
Mennekes
Mercedes
MIT
Mitsubishi
Most Corporation
NASA
NGK Plugs
Nissan
Nvidia
Oak Ridge National Labs
Peugeot
Philips

Pico Technology Automotive
Pierburg
Pioneer Radio
Pixabay
Porsche
Protean Electric
Renesas
Ricardo
Rimac
Rolec
Rover Cars
Saab Media
SAE
Scandmec
Schaeffler
Shutterstock
SMSC
Snap-on Tools
Society of Motor
    Manufacturers and Traders
    (SMMT)
Sofanou
Solvay
Sun Electric
Tesla Motors
Texas Instruments
Thatcham Research
Thrust SSC Land Speed Team
Topdon UK
Toyota

Acknowledgements

Tracker

Tula

Unipart Group

Valeo

Varta

Vauxhall

VDO Instruments

Volkswagen

Volvo Cars

Volvo Trucks

Wikimedia

YASA

ZF

If we have used any information or mentioned a company name that is not listed here, please accept our apologies and let us know so it can be rectified as soon as possible.

CHAPTER 1

Introduction to electric vehicles

1.1 Types of electric vehicle

1.1.1 Introduction

Here is an interesting thought to start with: The history of electric vehicles began between 1832 and 1839 when Robert Anderson of Scotland built a prototype electric carriage. Gustave Trouvé's tricycle (1881) is acknowledged as the world's first electric car. Earlier than many would have expected, I'm sure!

Figure 1.2 Volkswagen ID7–700 km WLTP range (Source: Volkswagen Media)

1.1.2 Pure electric vehicles

Pure electric vehicles (PEVs) have a battery instead of a fuel tank and an electric motor instead of an internal combustion engine (ICE). They are also described as battery electric vehicles (BEVs) or just electric vehicles (EVs). EVs are powered only by electricity and are plugged in to be recharged. They do not produce any tailpipe emissions. Most now have a real-world range of 100–300 miles on a single charge.

Figure 1.1 Gustave Trouvé's tricycle – limited range (1881)

DOI: 10.1201/9781003431732-1

Examples include the Nissan LEAF, Volkswagen ID3/4, Jaguar iPace, Audi Q4 e-tron and Tesla Model 3.

Figure 1.3 Audi Q4 e-tron PEV
(Source: Audi AG)

1.1.3 Hybrid electric vehicles

Hybrid electric vehicles (HEVs) are powered by an internal combustion engine in combination with one or more electric motors that use energy stored in batteries. HEVs combine the benefits of good fuel economy and low tailpipe emissions with the power and range of conventional vehicles. The standard type of HEV cannot be plugged in and charged from an external source.

Examples include the Toyota Prius, Honda CRV and BMW 3 Series.

Figure 1.4 Honda CRV HEV
(Source: Honda Media)

1.1.4 Plug-in hybrid electric vehicles

Plug-in hybrid electric vehicles (PHEVs) have a battery, electric drive motor and an internal combustion engine (ICE). They can be driven using the ICE, the electric drive motor or a combination of the two. They can be recharged from an external power source. Typical PHEVs have a pure-electric range of 20 to 50 miles. Once the battery is discharged, journeys can continue in hybrid mode, meaning that there is no range limit.

PHEV examples include the Mitsubishi Outlander PHEV, BMW 330e and Volkswagen Golf GTE.

Range-extended electric vehicles (E-REVs) are a type of PHEV, but they have an internal combustion engine and generator on board to recharge the battery when it is discharged. An example was the early BMW i3 but these are not now available as new.

Figure 1.5 Volkswagen Golf GTE

1.1.5 How to recognise an EV

Externally, the easiest way to recognise an EV (or HEV or PHEV!) is knowing the make or model of the vehicle or. if not, then by the badging. Many include an 'E' in the name, e-tron, GTE or BMW 330e, for example. The other outside feature will be the charging port. This may be under what looks like a fuel filler flap or, in some cases, behind the manufacturer's badge.

Inside the car, the dashboard will be a little different and, for example, will show the word 'Ready' when switched on – this is the equivalent of having the engine running. There is also likely to be a battery meter showing the state of charge.

Under the bonnet, the most obvious feature will be bright orange cables. These are the ones that carry the very high-voltage. They are perfectly safe to touch, if they are not damaged, but best left to the experts.

Many pure EVs have a green stripe on the number plates. If all else fails when identifying a vehicle, we could look in the driver's handbook!

Figure 1.6 Audi Q4 Sportback 50 e-tron quattro (Source: Audi Media)

1.2 Safety

1.2.1 High-voltage

Clearly a key safety issue with electric vehicles, of any type, is high-voltage. This is usually 400 volts or more (the Porsche Taycan, for example, is 800 volts), domestic mains voltage is 230 volts and that is dangerous! The high-voltage components include the

▶ battery
▶ power controller
▶ motor

▶ air conditioning
▶ heater

All of these components if accident damaged or dismantled are capable of giving a fatal electric shock. Again, all are perfectly safe for normal use, so don't worry.

Figure 1.7 Danger sign

1.2.2 Quiet operation

One aspect of a moving EV is that it can be very quiet. It is important, therefore, to look and listen when cars could be moving in the workshops or showrooms as well as car parks or anywhere similar.

Many EVs now have a noise generator that plays sound through an external speaker to improve safety. Although, smaller EVs, like heavy quadricycles, do not, so it always pays to be careful when working around such quiet vehicles.

1.2.3 Engines

On some hybrid vehicles, the engine can start without warning. This happens when the clever electronic systems detect that the battery is becoming discharged – the engine starts to recharge it. This will not happen if the vehicle is switched off and the keys removed.

1.2.4 **Personal protective equipment**

When working on certain aspects of a high-voltage vehicle, technicians will use appropriate personal protective equipment (PPE). However, no PPE is required by owners and users of EVs.

1.2.5 **Work on electric vehicles**

When work is being carried out, it is important that you do not approach an electric vehicle if it is cordoned off or displaying HV warning signs unless specifically asked to by a qualified person. Equally, you should never attempt to move a damaged electric vehicle without permission.

A technician should not work on the high-voltage or highly magnetic components of an EV if they have a heart pacemaker or other medical device such as an insulin pump fitted. Normal service and repair operations are fine. If in doubt, check the manufacturer's guidance.

Unless you are an EV-qualified technician carrying out diagnostic work, you should unplug vehicle from the charge socket before commencing any operation. An example would be when washing or valeting the vehicle.

If you see or hear anything unusual when operating or working on an EV, such as smoke or popping/hissing noises from battery or elsewhere, you should report this immediately.

Figure 1.8 Always disconnect the charger before washing or working on a vehicle

1.2.6 **Emergencies**

As with all vehicle types, dangerous situations can arise. In the event of a fire, the normal workplace procedures must be followed. If a person suffers an electric shock, then the instructions and processes outlined on the 'Electric Shock' posters should be followed.

A high-voltage vehicle that has been flooded or even completely submerged is not a particular danger but should only be moved or worked on by a qualified person.

1.2.7 **EVs are perfectly safe!**

It is important to stress again here that for normal use, drivers and passengers are perfectly safe in an EV. As well as insulation and shields, there are very sophisticated safety features built in; for example, the high-voltage is shut down in an impact. Clearly, you should not interfere with any complex systems on a vehicle and the high-voltage, in particular, unless you are a trained technician.

1.3 **Driving**

1.3.1 **Driving an EV**

Driving an EV is similar for almost every model. Turn the key or, more likely, press the *Power* button on the dash to turn the car on (*Ready* mode). Select *Drive*, which will usually be with a traditional gear lever, like the one in an automatic car.

Some manufacturers have moved the drive selector to the dashboard. Twist the selector towards the windscreen to go forwards and towards you for reverse. Other EVs may have a rotary dial with gear markings around it. With these, just turn the dial to select forward or reverse.

Pure EVs do not have gears or a clutch pedal, so press the accelerator pedal to go and the brake pedal to stop. It is not possible to stall an EV.

EVs are very smooth and quiet and a pleasure to drive. The instant torque of the electric motors means good acceleration, and, in some cars, *amazing* acceleration, so take care not to go too fast! Cruising at about 60 mph with smooth, gentle acceleration and gentle braking is the most economical.

EVs use something called regenerative braking. This converts movement energy into electricity to recharge the batteries. Because of this, an EV may slow down more than you expect when you lift the accelerator pedal. However, you can use this to your advantage as it makes the drive smoother and produces less wear on the brakes.

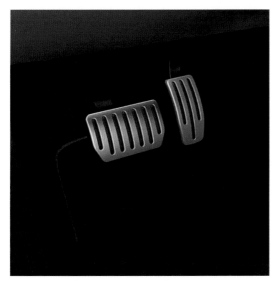

Figure 1.9 One pedal to go and one to stop! (Source: Tesla)

1.3.2 **Range**

The range of an EV is how far it can travel on a full charge. In other words, starting with the battery at 100% and driving it until the battery is completely discharged.

Vehicle range depends on a number of factors, including battery size, vehicle size, aerodynamics, battery heating efficiency and outside temperature. It is entirely normal for an electric vehicle to travel more miles on a charge on a hot summer day than a cold day in winter. It also depends very much on driving style (as it does for an ICE vehicle).

1.3.3 **Range anxiety**

Range anxiety is the fear that an electric vehicle will not have enough battery charge to reach its destination, leaving the motorist stranded. This anxiety is particularly prominent when considering long-distance travel, driving on roads where charging points could be few and far between or if someone is new to the technology. However, the charging infrastructure is improving every day.

The other key improvement is the actual range a vehicle will now achieve. These examples are based on standard test conditions, called the worldwide harmonised light vehicles test procedure (WLTP), but even if you reduce them by 25% to make them more 'real world', they are still good:

- ▶ Tesla Model 3 Long Range: 379 miles range
- ▶ Kia e-Niro Long Range: 283 miles range
- ▶ DS 3 Crossback: 199 miles range
- ▶ MINI Cooper S 1: 144 miles range

Just like with an ICE vehicle, it is important to choose a car that matches the user's needs. Interestingly, the average daily return journey length in the UK and EU is about 20 miles.

Figure 1.10 Kia e-Niro (Source: Kia Media)

1.3.4 **Other advantages**

While the initial upfront purchase price of an electric or plug-in hybrid vehicle can be higher, this can be offset by reduced running costs:

▶ A full charge in a pure electric vehicle will typically cost about half of the equivalent in fuel when owners charge at home and have access to an off-peak electricity tariff.

▶ There are fewer mechanical components in an electric vehicle compared with conventional vehicles, which often results in lower servicing and maintenance costs.

▶ There is a lower vehicle excise duty (VED).

▶ EVs offer free or cheaper access to low emission and congestion zones.

▶ There is free parking for electric vehicles available in some towns and cities.

▶ There are currently incentives for company cars to switch to greener travel.

1.4 **Charging**

1.4.1 **Location of charging ports**

The charging port is where you plug in the cable that connects to the mains or to a dedicated charger. It may be under what looks like a fuel filler flap or, in some cases, behind the manufacturer's badge. There are a few different types, and they can be at the front, rear or on the side of the vehicle.

1.4.2 **Charging methods**

There are essentially three ways to charge an EV:

1. Trickle charge: This is the slowest method of charging your EV at home, using a standard 230 V plug. It is only recommended in emergencies.

2. AC charge: Having a wall box installed lets you charge 3–4 times faster using AC domestic charging. AC public charging is also widely available and is popular for workplaces too.

3. DC charge: This is the fastest way to charge an EV, but older EVs and PHEVs do not have this option. These public DC fast-charging stations usually have power rated at 50 kW and above. With this method, for example, you can top up a 60 kWh battery from 20% to 80% in about 40 minutes. There are now also some ultra-fast charging stations that provide 150 kW and 350 kWh chargers will become available in the future.

> **Key Fact**
> DC chargers can deliver power faster as they have large inverters in the charging station, and this bypasses the vehicle's on-board inverter, delivering power straight to the battery.

1.4.3 **Time to recharge**

A typical recharge, particularly when travelling, is from 20% to 80%. Table 1.1 shows average times of the different methods for a 100 kWh battery (which is a big battery):

However, even though very fast charging is possible, Figure 1.11 offers some very important advice for prolonging the life of a battery.

1.5 **Test cycle and emissions**

1.5.1 **Carbon dioxide**

Electric vehicles have zero emissions at the point of use, so-called 'tank-to-wheel', when powered only by the battery. The 'well-to-wheel' analysis includes the CO_2 emissions during electricity generation, which depend on the current mixture of fuels used to make the electricity for the grid. To make a correct comparison with emissions from all cars, it is necessary to use the 'well-to-wheel' figure, which includes the CO_2 emissions during production, refining and distribution of petrol/diesel.

Table 1.1 Typical charging times

Device	Power (kW)	Approximate time (0% to 100%)	Approximate time (20% to 80%)
Domestic EV charger	3	33 hours	20 hours
Domestic wall box charger	7	14 hours	9 hours
AC fast charger	22	4 hours 30 mins	3 hours
DC fast charger	60	1 hour 40 mins	1 hour
DC ultra-fast charger	150	40 mins	24 mins
DC fastest charger*	350	16 mins	10 mins

*becoming available at the time of writing but not in common use and not suitable for many cars

Store cold, use hot   Charge slowly   Store half full

Figure 1.11 Prolonging the life of a battery (in your phone and car)

Key Fact

Electric vehicles have zero emissions at the point of use, so-called 'tank-to-wheel'.

Electricity production continues to decarbonise because of reduced reliance on oil and coal, so the overall emission figure for running an EV will drop further. Tailpipe emissions also include oxides of nitrogen (NO_x) and particulate matter (tiny particles of solid or liquid matter suspended in a gas or liquid) that contribute to air pollution. This is why any vehicle operating only on battery power will play a significant role in improving local air quality.

1.5.2 Worldwide harmonised light vehicles test procedure

The worldwide harmonised light vehicles test procedure (WLTP) is a global standard for determining levels of pollutants and CO_2 emissions, fuel or energy consumption and electric range (Figure 1.12). It is used for passenger cars and light commercial vans. Experts from the EU, Japan and India, under guidelines of the UNECE World Forum for Harmonization of Vehicle Regulations, developed the standard. It was released in 2015.

Definition

WLTP: Worldwide harmonised light vehicles test procedure.

Like all previous test cycles, the WLTP has its drawbacks, but it is a good attempt to make the test more realistic. The key thing is that a standardised test allows vehicles to be compared accurately, even if the figures differ from real-world driving. It includes strict guidance regarding conditions of dynamometer tests and road load (motion resistance), gear changing, total car weight (by including

7

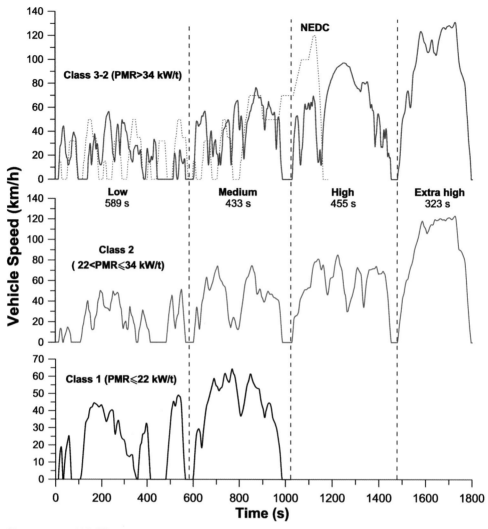

Figure 1.12 WLTP test cycles

optional equipment, cargo and passengers), fuel quality, ambient temperature and tyres and their pressure. Three different cycles are used depending on vehicle class defined by power/weight ratio PWr in kW/tonne (rated engine power/kerb weight):

▶ Class 1 – low power vehicles with PWr ≤ 22
▶ Class 2 – vehicles with 22 < PWr ≤ 34
▶ Class 3 – high-power vehicles with PWr > 34

Most modern cars, light vans and busses have a power/weight ratio of 40–100 kW/t, so they belong to Class 3, but some can be in Class

2. In each class, there are several driving tests designed to represent real-world vehicle operation on urban roads, extra-urban roads, motorways and freeways.

Key Fact

Most modern cars, light vans and busses have a power/weight ratio of 40–100 kW/t, so belong to WLTP Class 3.

The WLTC driving cycle for a Class 3 vehicle is divided into four parts:

1. Low
2. Medium
3. High
4. Extra-high speed

If Vmax < 135 km/h, the Extra-high speed part is replaced with a Low speed part.

1.5.3 Life cycle assessment

To determine the carbon footprint of a vehicle, its life is divided into three phases (Figure 1.13):

1. Production
2. Use
3. Recycling

For most manufacturers (Volkswagen is the example used here), the carbon footprint of the electrically propelled variants is already better than that of the corresponding vehicles with internal combustion engines. The same vehicle model but with different powertrains is used for this comparison.

> **Key Fact**
>
> To determine the carbon footprint of a vehicle, its life is divided into three phases:

▶ *Production*
▶ *Use*
▶ *Recycling*

Electric vehicles offer a higher CO_2 saving potential in all phases of the product cycle. Furthermore, it is of crucial importance for CO_2 emissions whether the propulsion energy is generated from fossil or regenerative sources.

As an example from 2019, the Golf TDI (Diesel) emits 140 g CO_2/km on average over its entire life cycle, while the e-Golf reaches 119 g CO_2/km.

A vehicle with an internal combustion engine creates most of its emissions during the use phase, that is, in the supply chain of the fossil fuel and the combustion. During this phase, the diesel reaches 111 g CO_2/km. A corresponding vehicle with electric drive emits only 62 g CO_2/km during this phase, the result of energy generation and supply. By contrast, most emissions from the battery-powered electric vehicle are generated in the production phase. According to life cycle assessment (LCA), a diesel vehicle generates 29 g CO_2/km, while a comparable e-vehicle generates 57 g CO_2/km.

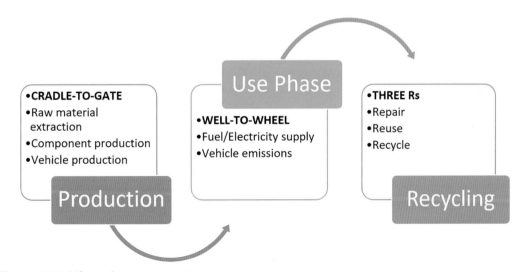

Figure 1.13 Life cycle assessment

Battery production and the complex extraction of raw materials are responsible for the additional CO_2 for e-vehicles. These emissions account for almost half of the CO_2 emissions of the entire life cycle. During the use phase, CO_2 emissions depend on the sources of energy production. They decrease as more regenerative energies are available.

Life cycle assessment (LCA) is an intricate, complex and internationally standardised procedure to research the ecological balance sheet of vehicles. Among other things, carbon dioxide emissions are investigated during all production stages of the automobile:

▶ The emissions generated by the extraction of raw materials, the production of components and the assembly of the car are included in the production phase.
▶ The use phase includes both the emissions of the fuel and electricity supply and especially those of vehicle operation over 200,000 km.
▶ Recycling evaluates dismantling and the potential savings of recycling.

Findings from the LCA mean that Volkswagen and other manufacturers can derive additional emission-reducing measures for life cycle engineering (LCE) and, specifically, optimise the CO_2 balance.[1]

1.5.4 Intelligent programming

The intelligent programming of autonomous electric vehicles can result in energy savings. Most BEVs allow the driver a choice of driving styles, for example, economy, normal and sport. It is several years away yet, but an autonomous Level 4 to 5 vehicle can result in better miles per kW/h than a human driver. Map data is used to determine the route and details from the route such as speed limits, required turns, intersections and time factors are preprogrammed.

Driving dynamics, drag of the vehicle and changes in the battery state of charge (SOC) are also considered. SOC affects everything from route selection to availability and usability of regenerative braking. Other factors that change during the route include vehicle-to-infrastructure transmissions (traffic control devices, for example), sensor data and camera readings from on-board devices and even changes in the weather.

The most efficient way to decelerate is by coasting because there is no active reduction in speed and no waste of energy. However, unless the battery SOC is less than 90%, there will be little or no energy recovery from regeneration. Driving in a large town or city makes coasting difficult. An energy-efficient urban acceleration-and-braking approach is therefore used. If conditions don't permit the precalculated drive, the vehicle is operated by slow, light acceleration and gentle braking until opportunities for the preset strategies reappear.

> **Key Fact**
> The most efficient way to decelerate is by coasting.

Savings of battery power of over 10% are potentially achievable and, in some cases, up to 50%, when compared to sportier driving styles.

1.6 Fear, uncertainty and doubt

1.6.1 Introduction

FUD is an acronym for fear, uncertainty and doubt, and there is a lot of it about in relation to electric and hybrid vehicles. This is perfectly understandable and often happens when things seem to be changing quickly. However, it can also be used as a tool for misinformation – and this is definitely the case with electric vehicles.

Our personal view, and that of almost all motor manufacturers, is that EVs are part of the future, not only for their role in reducing emissions and global warming but because they are much nicer to drive than ICE vehicles for personal use.

An electric car uses energy more efficiently. A fossil burner wastes 70–75% of the energy it burns, and fossil fuel is extracted, transported and refined before use, releasing vast amounts of CO_2 never included in 'tailpipe' emissions.

Even hydrogen fuel cell cars are much less efficient than a pure EV. This is because of the energy costs in creating, transporting and converting the hydrogen back to electricity. Hydrogen as a fuel has its advantages, particularly in plant and heavy-duty vehicles.

Using the correct fuel source for the correct application is an environmentally responsible decision that many will have to make in order to reduce emissions that are harmful.

Figure 1.14 At the bottom of this mountain road in the French Alps, my EV had five more miles of range than it did at the top

Figure 1.15 Electric London taxi – the driver of one of these said he saves £5,000 per year on fuel costs compared to the older diesel version

1.6.2 EV myths

Electric vehicles have a poor range because they run out of electricity after a short distance

The average length of a commuter trip by car/van across England and Wales is about 16 kilometres (10 miles). On average, European workers spend 1 hour and 24 minutes a day commuting, travelling 28.56 km (17.7 miles) in total. In terms of distance undertaken, about one-fifth of Europeans (23.3%) travel, on average, further than 40 km or even more each day, while 36.2% travel fewer than 10 km. Americans drive an average of 40 miles a day.

Even the shortest-range electric vehicles claim to cover more than twice these distances before being recharged. However, a value of two-thirds of the claimed miles is a more realistic distance in the real world. Even if you halve the following WLTP figures (for example, for a really cold day or poor driving style), then for most commuting journeys, the vehicles are fine:

- ► Tesla Model 3 Long Range: 379 miles/610 km range
- ► Kia e-Niro Long Range: 283 miles/455 km range
- ► DS 3 Crossback: 199 miles/320 km range
- ► MINI Cooper S 1: 144 miles/232 km range
- ► Volkswagen ID7: 435 miles/700 km range

Electric vehicles are slow

The electric motor produces 100% of its torque instantly. This is the same at all speeds – in other words, a flat torque curve. One version of the Tesla Model S, when in *Ludicrous* mode, is one of the quickest production cars in the world – it has a 0–60 mph time of about 2.5 seconds.

Electric vehicles are expensive

It is often claimed that EVs are more expensive than ICE vehicles – or are they? The cost of the battery is a defining factor here, but this is falling and is expected to drop further in the

11

next few years. It is important to consider the longer-term cost. Over a period of time, even though some EVs are more expensive, the running costs are significantly less, so they can work out cheaper overall.

Electric vehicles are not safe

In crash tests, EVs must meet the same, if not more, stringent standards as ICE vehicles – and they do! The high-voltage employed by EVs is definitely dangerous if worked on by untrained technicians, but this is now covered by schemes such as IMI TechSafe (see page 16). The high-voltages pose no risk to drivers and passengers, even in a crash. A recent study on vehicle fires concluded that the frequency and severity of fires and explosions from lithium-ion battery systems are comparable to, or less than, ICE cars.

Electric vehicles produce more emissions than ICE vehicles

EVs convert at least 75% of the chemical energy in the batteries to mechanical energy at the wheels. ICE vehicles only convert about 25% of the energy stored in petrol/gasoline. EVs do not produce tailpipe emission/pollutants. They do produce emissions at the electricity generation plants, but the conversion here is still more efficient than an ICE. However, there is a significant trend towards greener power generation, so EVs are becoming cleaner every day. It is also important to consider the full life cycle of a vehicle – see Section 1.5.3 for more details.

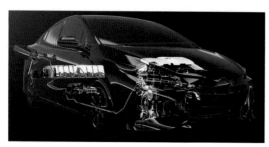

Figure 1.16 Toyota Prius plug-in hybrid (Source: Toyota Media)

Electric cars are more expensive to service and repair

EVs do need to be checked and serviced regularly like all complex machinery. However, there are far fewer moving parts to eventually fail. Comparing like-for-like vehicles at main dealers, servicing costs are less for EVs.

There are not enough public charging points

Most EV charging is done at home or at work. However, in the UK (as of December 2019) the total number of locations which have a public charging point installed was 10,343, the number of devices at those locations was 16,495 and the total number of connectors within these devices was 28,541. The USA had over 20,000 electric car charging stations with more than 68,800 connectors. In 2021, the EU had an estimated 375,000 charging stations. In just over two years (April 2023), these numbers increased to:

▶ 40,150 charging points located throughout the UK[2]
▶ 130,000 charging stations in the USA[3]
▶ 500,000+ charging points located across the EU[4]

Batteries only last a few years

Most vehicle manufacturers warranty their battery packs for at least 8 years or 100,000 miles. Several reports have shown that Nissan

Figure 1.17 Toyota Prius presenting for Ministry of Transport (MOT) test

LEAF models still had 75% of their battery capacity after 120,000 miles. One Tesla owner claims to have 90% of their car's battery capacity, even after 200,000 miles.

We personally know of several 1997/8 Toyota Prius cars still running well on their original batteries. Up to 99% of EV batteries can be repaired, remanufactured, reused or recycled.

The power grid will not be able to handle the extra load

Power grids across the UK, EU and USA can handle the expected growth in EVs for the foreseeable future without any significant change. This is because most EVs tend to be charged at night during off-peak hours and when demand is usually lowest. To balance the grid, it will be necessary to agree to methods of timing, such as competitive pricing for charging EVs during certain hours, but capacity is not the problem.

Reserves of rare earth metals used in batteries and motors will run out

We first need to define the difference here between a resource and reserves of that resource. A resource is how much of a geologic commodity exists. It includes discovered and undiscovered amounts; therefore, it is a calculated estimate. Reserves are the amount of a resource that have been discovered. For this reason, they have a known quantity and are financially viable for extraction. This viability changes depending on technologies and demand. In other words, as demand for certain rare earth metals increases so do the available reserves. Other technological developments are also taking place. For example, motors now contain significantly less rare-earth element.

1.6.3 EV speed

The very first production version of the Rimac Nevera (Figure 1.18), the world's fastest accelerating production car, is ready to take to

the streets. With 1,914 hp produced by four electric motors, the electric hypercar reaches 0–60 mph in just 1.85 seconds and to 100 mph in 4.3 seconds.

After an extensive five-year journey of development and rigorous testing, involving three generations of powertrain technology, 18 prototypes, 45 physical crash tests and over 1.6 million collective hours of research and development, the Rimac Nevera has reached a significant milestone as the first full production version rolls off Rimac's newly established production line (2022/3). The hand-building process of the Nevera will now continue at a steady pace, with up to 50 units being crafted annually.

As we speak, the initial batch of cars, eagerly awaited by customers, is already in the process of assembly. Each Nevera undergoes a meticulous five-week final assembly line procedure, with various components and systems manufactured months in advance at Rimac facilities. Soon, these remarkable vehicles will be delivered to destinations worldwide through Rimac's extensive network of 25 official dealer partners, spanning the United States, Europe, the Middle East and Asia.

The Rimac Nevera stands as a testament to innovation, being the world's first all-electric hypercar developed entirely from scratch.

Figure 1.18 The Rimac Nevera is the world's fastest accelerating production car (Source: Rimac)

Figure 1.19 Not quite as fast as the Rimac Nevera but this Citroen could be the future of local transport

With a relentless pursuit of high-performance targets, Rimac has meticulously crafted and engineered most of the key systems in-house. Custom development was the norm for all major components, including an entirely new generation battery system, inverter, gearbox, motor, control systems, infotainment and more. Over a span of two years, the powertrain underwent three complete redesigns, showcasing the dedication and commitment of the team behind this ground-breaking achievement.

Notes

1   Source: Volkswagen Media.
2   https://www.gov.uk/government/statistics/ electric-vehicle-charging-device-statistics-october-2022/electric-vehicle-charging-device-statistics-october-2022
3   https://usafacts.org/articles/how-many-electric-vehicle-charging-stations-are-there-in-the-us/
4   https://www.statista.com/statistics/955443/ number-of-electric-vehicle-charging-stations-in-europe/

CHAPTER 2

Safe working, tools and hazard management

2.1 General safety precautions

2.1.1 Introduction

Safe working practices in relation to all automotive systems are essential, for your safety as well as that of others. When working on high-voltage systems, it is even more important to know what you are doing. However, you only have to follow two rules to be safe:

▶ Use your common sense – don't fool about.
▶ If in doubt, seek help.

The following section lists some particular risks when working with electricity or electrical systems, together with suggestions for reducing them. This is known as risk assessment.

> **Definition**
> Risk assessment: a systematic process of evaluating the potential risks that may be involved in an activity or undertaking.

2.1.2 Safety

Electric vehicles (pure or hybrid) use high-voltage batteries so that energy can be delivered to a drive motor or returned to a battery pack in a

very short time. The Honda Insight system, for example, uses a 144 V battery module to store regenerated energy. The Toyota Prius originally used a 273.6 V battery pack, but this was changed in 2004 to a 201.6 V pack. Voltages of 400 V are now common and some up to 700 V, so clearly, there are electrical safety issues when working with these vehicles.

EV batteries and motors have high electrical and magnetic potential that can severely injure or kill if not handled correctly. It is essential that you take note of all the warnings and recommended safety measures outlined by manufacturers and in this resource. Any person with a heart pacemaker or any other electronic medical device should not work on an EV motor since the magnetic effects could be dangerous. In addition,

Figure 2.1 Volvo hybrid car (Source: Volvo Media)

DOI: 10.1201/9781003431732-2

other medical devices such as intravenous insulin injectors or meters can be affected.

Safety First

EV batteries and motors have high electrical and magnetic potential that can severely injure or kill if not handled correctly.

Most of the high-voltage components are combined in a power unit. This is often located behind the rear seats or under the luggage compartment floor (or the whole floor in a Tesla). The unit is a metal box that is completely closed with bolts. A battery module switch, if used, may be located under a small secure cover on the power unit. The electric motor is located between the engine and the transmission or as part of the transmission on a hybrid, or on a pure EV, it is the main driving component. A few vehicles use wheel motors too.

The electrical energy is conducted to or from the motor by thick orange wires. If these wires have to be disconnected, switch off or de-energise the high-voltage system. This will prevent the risk of electric shock or short circuit of the high-voltage system.

Safety First

High-voltage wires are always orange.

NOTE: Always follow the manufacturer's instructions – it is not possible to outline all variations here.

Figure 2.2 EV battery and drive components (Source: Porsche Media)

2.1.3 IMI TechSafe™

IMI TechSafe™ is professional recognition within the IMI Professional Register. It identifies a member's professionalism and safe working practice in the field of electric vehicles (EV) and other safety-critical vehicle systems such as autonomous or advanced driver assistance systems (ADAS). As well as proof of competence through the achievement of nationally recognised qualifications or IMI accreditations, having IMI TechSafe recognition means that the member keeps up-to-date through mandatory requirements for continuous professional development (CPD). These CPD requirements are decided and agreed upon by an IMI industry sector advisory group and are reviewed on a regular basis.

Employers have a responsibility and duty of care to ensure their staff are competent to work on electric vehicles (EV) and that they meet the requirements of the *Electricity at Work Regulations* 1989. The repair of damaged electric vehicles and those with ADAS is covered within the British Standard, BS10125. Most insurance companies (work providers) support this by only giving work to businesses that meet the standard. Vehicle manufacturers set out repair methods and processes from a servicing and maintenance point of view as well as for the repair of damaged vehicles. These must be followed to ensure their own safety, the safety/ roadworthiness of the vehicle and the safety of the driver and other road users or pedestrians.

IMI TechSafe recognition gives an employee the ability to easily demonstrate to colleagues, customers and professionals that they're fully qualified to be working on EV vehicles and/or vehicles with ADAS

Figure 2.3 Why do I need to be IMI TechSafe?

technologies. It is also seen as a badge of honour for those in the sector who hold it and can lead to increased employability as employers are searching for EV technicians to join their work forces.

There are a few steps to take to gain IMI TechSafe recognition within the IMI Professional Register:

▶ gain a qualifying achievement, for example, an IMI nationally recognised qualification or IMI accreditation
▶ join the IMI membership community
▶ take your place on the IMI's Professional Register
▶ gain IMI TechSafe recognition

A vital part of the Professional Register and IMI TechSafe is ensuring those on it remain up-to-date with their knowledge, skills and competency. This is why CPD is a requirement to remain in IMI TechSafe recognition.

2.1.4 General safety guidance

Before maintenance
▶ Turn off the ignition switch and remove the key.
▶ Switch off the battery module switch or de-energise the system.
▶ Wait for five minutes before performing any maintenance procedures on the system. This allows any storage capacitors to be discharged.

During maintenance
▶ Always wear insulating gloves.
▶ Always use insulated tools when performing service procedures to the high-voltage system. This precaution will prevent accidental short circuits.

Interruptions
When maintenance procedures have to be interrupted while some high-voltage components are uncovered or disassembled, make sure that

▶ the ignition is turned off and the key is removed

Figure 2.4 High-voltage gloves (Class 0, short style)

▶ the battery module switch is switched off
▶ no untrained persons have access to that area and prevent any unintended touching of the components.

After maintenance

Before switching on or re-energising the battery module after repairs have been completed, make sure that

▶ all terminals have been tightened to the specified torque
▶ no high-voltage wires or terminals have been damaged or shorted to the body
▶ the insulation resistance between each high-voltage terminal of the part you disassembled and the vehicle's body has been checked

Working on electric and hybrid vehicles is not dangerous *if* the previous guidelines and the manufacturer's procedures are followed. Before starting work, check the latest information – DON'T take chances. Dying from an electric shock is not funny.

Crash safety: Electric vehicles are tested to the same high standards as other vehicles currently on UK roads. In February 2011, the first pure electric car was assessed and passed the renowned Euro New Car Assessment Programs (NCAP) test.

Figure 2.5 High-voltage cables are always orange (Source: Volkswagen Media)

Pedestrian safety: The quietness of EVs is a benefit but can pose a threat to sight- and hearing-impaired people, particularly at low speeds. Having seen a vehicle, pedestrians are capable of reacting to avoid an accident at vehicle speeds up to 15 mph. However, research found that tyre noise will alert pedestrians to a vehicle's presence at speeds greater than 12.4 mph.

2.1.5 General risks and their reduction

Table 2.1 lists some identified risks involved with working on ALL vehicles. The table is by no means exhaustive but serves as a good guide.

> **Safety First**
>
> Electric vehicles are tested to the same high standards as other vehicles.

> **Safety First**
>
> Do not touch any electric circuit that is greater than the standard 12 V or 24 V.

Table 2.1 Risks and their reduction

Identified risk	Reducing the risk
Electric shock 1	Voltages and the potential for electric shock when working on an EV mean a high risk level – see Section 2.2 for more details.
Electric shock 2	Ignition high tension (HT) is the most likely place to suffer a shock when working on an ICE vehicle; up to 40,000 V is quite normal. Use insulated tools if it is necessary to work on HT circuits with the engine running. Note that high-voltages are also present on circuits containing windings due to back EMF as they are switched off; a few hundred volts is common. Mains-supplied power tools and their leads should be in good condition, and using an earth leakage trip is highly recommended. Only work on HEV and EVs if trained in the high-voltage systems.
Battery acid	Sulfuric acid is corrosive, so always use good personal protective equipment. In this case, overalls and, if necessary, rubber gloves. A rubber apron is ideal as are goggles if working with batteries a lot.
Raising or lifting vehicles	Apply brakes and/or chock the wheels when raising a vehicle on a jack or drive-on lift. Only jack under substantial chassis and suspension structures. Use axle stands in case the jack fails.
Running engines	Do not wear loose clothing; good overalls are ideal. Keep the keys in your possession when working on an engine to prevent others starting it. Take extra care if working near running drive belts.
Exhaust gases	Suitable extraction must be used if the engine is running indoors. Remember it is not just the carbon monoxide that might make you ill or even kill you; other exhaust components could cause asthma or even cancer.
Moving loads	Only lift what is comfortable for you; ask for help, if necessary, and/or use lifting equipment. As a general guide, do not lift on your own if it feels too heavy!
Short circuits	Use a jump lead with an in-line fuse to prevent damage due to a short when testing. Disconnect the battery (earth lead off first and back on last) if any danger of a short exists. A very high current can flow from a vehicle battery; it will burn you as well as the vehicle.
Fire	Do not smoke when working on a vehicle. Fuel leaks must be attended to immediately. Remember the triangle of fire: Heat – Fuel – Oxygen. Don't let the three sides come together.
Skin problems	Use a good barrier cream and/or latex gloves. Wash skin and clothes regularly.

2.2 High-voltage safety precautions

2.2.1 Introduction

In this section we will consider the differences between AC and DC as well as high and low voltages. This can be confusing, so let's keep it very simple to start with:

The voltages (AC or DC) used on electric vehicles can kill, have killed and will kill again.

Follow all safety procedures and do not touch any electric circuit that is greater than the standard 12 V or 24 V that we are used to, and you will be fine!

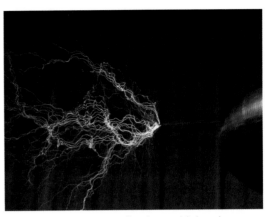

Figure 2.6 Plasma trails due to high-voltage

2.2.2 Low and high-voltage

Low-voltage is a relative term, the definition varying by context. Different definitions are used in electric power transmission and distribution and in the electronics industry. Electrical safety codes define low-voltage circuits, which are exempt from the protection required at higher voltages. These definitions vary by country and specific code. The International Electrotechnical Commission (IEC) define voltages as in Table 2.2.

> **Definition**
> IEC: International Electrotechnical Commission.

2.2.3 Defining high-voltage

For EVs, the United Nations (UN) document: Addendum 99: Regulation No. 100 Revision 3, states: 'High-voltage' means the classification of an electric component or circuit if its working voltage is > 60 V and ≤ 1500 V DC or > 30 V and ≤ 1000 V AC root mean square (rms).

> **Safety First**
> For EVs, DC voltages between 60 V and 1500 V are referred to as 'high-voltage'.

2.2.4 Personal protective equipment (PPE)

In addition to the normal automotive-related PPE, the following are also recommended for work on high-voltage systems:

- ▶ overalls with non-conductive fasteners
- ▶ electrical protection gloves
- ▶ protective footwear with rubberised soles and non-metallic protective toe caps
- ▶ goggles/face shield (when necessary)

Table 2.2 IEC voltages

IEC voltage range	AC	DC	Defining risk
High-voltage (supply system)	>1000 Vrms*	>1500 V	Electrical arcing
Low-voltage (supply system)	50–1000 Vrms	120–1500V	Electrical shock
Extra low-voltage (supply system)	<50 Vrms	<120 V	Low risk

*The root mean square (rms) is a value characteristic of a continuously varying quantity, such as an AC electric current. This is the effective value in the sense of the value of the direct current that would produce the same power dissipation in a resistive load.

Safety First

Electrical safety gloves are NOT the same as general working gloves.

Personal protective equipment (PPE) is essential for EV work. Electrical safety gloves are categorised by the level of voltage protection they provide. The voltage breakdown is as follows for gloves appropriate to EV work:

▶ Class 00 is rated at a maximum use voltage of 500 V AC/750 V DC and proof-tested to 2,500 V AC/10,000 V DC
▶ Class 0 is rated at a maximum use voltage of 1,000 V AC/1,500 V DC and proof-tested to 5,000 V AC/20,000 V DC
▶ Class 1 is rated at a maximum use voltage of 7,500 V AC/11.250 V DC and proof-tested to 10,000 V AC/40,000 V DC.

Class 00 is adequate, but Class 0 is recommended because EV voltages are tending to increase.

Gloves should be inspected for tears, holes, cuts and other defects before each use. Also, check for any swelling, which can be caused by contamination with petroleum products. An air test should be performed along with inspections for insulating gloves. The glove is filled with air and then checked for leakage.

The gloves can get quite sweaty, so cotton lining gloves may make them more comfortable to wear. Armoured outer gloves can also protect electrical protection gloves, but wearing three pairs of gloves can make some work difficult! Protection from hazards should always take priority, and remember that electrical protection gloves may have a life of about six months, so it is important to check the glove manufacturer's guidance.

If the gloves show any signs of defects, they should be taken out of service.

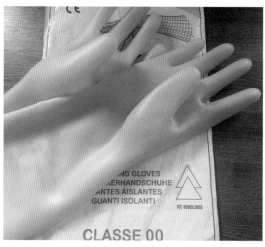

Figure 2.7 Class 00 electrical gloves

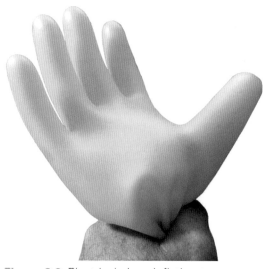

Figure 2.8 Electrical glove inflation test to check for cuts and damage

Safety First

▶ *Gloves should be inspected for tears, holes, cuts and other defects before each use.*
▶ *Make sure the gloves are in date.*

2.2.5 High-energy cables and components

Electric vehicles use high-voltage batteries so that energy can be delivered to a drive

Figure 2.9 Orange high-voltage cables

Figure 2.11 Warning labels

Figure 2.10 Danger sticker

Figure 2.12 General warning sticker

motor or returned to a battery pack efficiently in a very short time. It is important to be able to correctly identify high-energy cabling and associated components. This is done by

▶ colouring
▶ warning symbols
▶ warning signs

The following pictures show the location of high-voltage components and wires (orange) together with some warning stickers.

2.2.6 **AC electric shock**

When an AC current exceeding 30 mA passes through a part of a human body, the person concerned is in serious danger if the current is not interrupted in a very short time. The protection of persons against electric shock must be provided in conformity with

appropriate national standards, statutory regulations, codes of practice, official guides and circulars.

> **Safety First**
> When an AC current exceeding 30 mA passes through a part of a human body, the person concerned is in serious danger.

An electric shock is the physical effect of an electric current through the human body. It affects the muscular, circulatory and respiratory functions and sometimes results in serious burns. The degree of danger for the victim is a function of the size of the current, the parts of the body through which the current passes and the duration of current flow.

21

IEC publication 60479–1 defines four zones of current-magnitude/time-duration, in each of which the pathophysiological effects are described. Any person coming into contact with live metal risks an electric shock.

Curve C_1 in Figure 2.13 shows that when a current greater than 30 mA passes through a human body from one hand to the feet, the person concerned is likely to be killed unless the current is interrupted in a relatively short time. This is the real benefit of modern residual current circuit breakers because they can trip before serious injury or death!

Here are the abstracts from some appropriate international standards:

IEC 60479–1 Abstract: For a given current path through the human body, the danger to persons depends mainly on the magnitude and duration of the current flow. However, the time/current zones specified in this publication are, in many cases, not directly applicable in practice for designing measures of protection against electrical shock. The necessary criterion is the admissible limit of touch voltage (i.e. the product of the current through the body called touch current and the body impedance) as a function of time. The relationship between current and voltage is not linear because the impedance of the human body varies with the touch voltage, and data on this relationship is therefore required. The different parts of the human body (such as the skin, blood, muscles, other tissues and joints) present to the electric current a certain impedance composed of resistive and capacitive components. The values of body impedance depend on a number of factors and, in particular, on current path, on touch voltage, duration of current flow, frequency, degree of moisture of the skin, surface area of contact, pressure exerted and temperature. The impedance values indicated in this technical specification result from a close examination of the experimental results available from measurements carried out principally on corpses and on some living persons. This technical specification has the status of a basic safety publication in accordance with IEC Guide 104.[1]

IEC 60479–2 Abstract: [This] technical specification describes the effects on the human body when a sinusoidal alternating current in the frequency range above 100 Hz passes through it. The effects of current passing through the human body for:

▶ alternating sinusoidal current with DC components
▶ alternating sinusoidal current with phase control
▶ alternating sinusoidal current with multicycle control are given but are only deemed applicable for alternating current frequencies from 15 Hz up to 100 Hz.

This standard furthermore describes the effects of current passing through the human body in the form of single unidirectional rectangular impulses, sinusoidal impulses and impulses resulting from capacitor discharges. The values specified are deemed to be applicable for impulse durations from 0.1 ms up to and including 10 ms. [. . .] This standard only considers conducted current resulting from the direct application of a source of current to the body, as does IEC 60479–1 and IEC 60479–3. It does not consider current induced within the body caused by its exposure to an external electromagnetic field. This third edition cancels and replaces the second edition, published in 1987, and constitutes a technical revision. The major changes with regard to the previous edition are as follows:

▶ the report has been completed with additional information on effects of current passing through the human body for alternating sinusoidal current with DC components, alternating sinusoidal current with phase control, alternating sinusoidal current with multicycle control in the frequency range from 15 Hz up to 100 Hz

- an estimation of the equivalent current threshold for mixed frequencies
- the effect of repeated pulses (bursts) of current on the threshold of ventricular fibrillation
- effects of electric current through the immersed human body.[2]

How AC affects the body depends very much on the frequency. Low-frequency AC is used in US (60 Hz) and European (50 Hz) households and industries. This can be more dangerous than high-frequency AC and is three to five times more dangerous than DC of the same voltage and amperage.

This is because AC has a greater tendency to throw the heart into a condition of

fibrillation, whereas DC tends to just make the heart stop. Once the shock current is halted, a stopped heart has a better chance of regaining a normal beat pattern than a fibrillating heart. This is why defibrillating equipment is used by first aiders. The jolt of current supplied by the defibrillator is DC, which stops fibrillation and gives the heart a chance to recover.

Low-frequency AC produces extended muscle contraction, which may freeze a person's hand to the electrical source, therefore prolonging exposure. DC is most likely to cause a single convulsive contraction, which often forces the victim away from the source.

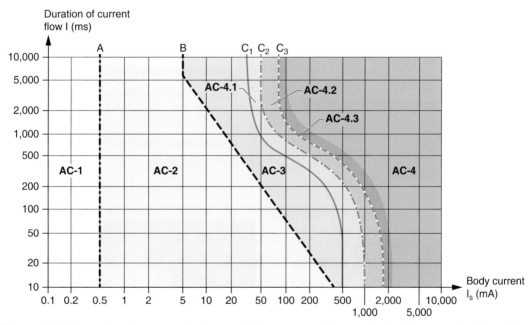

Figure 2.13 Zones time/current of effects of AC current on human body when passing from left hand to feet: AC-1 zone: imperceptible; AC-2 zone: perceptible; AC-3 zone: reversible effects, muscular contraction; AC-4 zone: possibility of irreversible effects; AC-4.1 zone: up to 5% probability of heart fibrillation; AC-4.2 zone: up to 50% probability of heart fibrillation; AC-4.3 zone: more than 50% probability of heart fibrillation. A curve: threshold of perception of current; B curve: threshold of muscular reactions; C_1 curve: threshold of 0% probability of ventricular fibrillation; C_2 curve: threshold of 5% probability of ventricular fibrillation; C_3 curve: threshold of 50% probability of ventricular fibrillation.
(Source: www.electrical-installation.org/enwiki/Electric_shock)

2.2.7 DC electric shock

The three basic factors that determine what kind of shock you experience when current passes through the body are

- size of the current
- duration
- frequency

Direct currents have zero frequency, as the current is constant. DC causes a single continuous contraction of the muscles compared with AC current, which will make a series of contractions depending on the frequency. In terms of fatalities, both kill but more milliamps are required of DC current than AC current at the same voltage.

> **Definition**
> Fibrillation: a condition when all the heart muscles start moving independently in a disorganised manner.

Either AC or DC currents can cause fibrillation of the heart at high enough levels. This typically takes place at 30 mA of AC (rms, 50–60 Hz) or 300–500 mA of DC.

Facts about electric shock:

- It is the magnitude of current and the duration that produces effect. That means a low-value current for a long duration can also be fatal. The current/time limit for a victim to survive at 500 mA is 0.2 seconds and at 50 mA is 2 seconds.
- The voltage of the electric supply is only important because it ascertains the magnitude of the current. As Current = Voltage/Resistance, the bodily resistance is an important factor. Sweaty or wet persons have a lower body resistance, so they can be fatally electrocuted at lower voltages.
- Let-go current is the highest current at which the subject can release a conductor. Above this limit, involuntary clasping of the conductor occurs: It is 22 mA in AC and 88 mA in DC.
- Severity of electric shock depends on body resistance, voltage, current, path of the current, area of contact and duration of contact
- Heating due to resistance can cause extensive and deep burns; damaging temperatures are reached in a few seconds.

An arc flash is the light and heat produced from an electric arc supplied with sufficient electrical energy to cause substantial damage, harm, fire or injury. Note that welding arcs can turn steel into a liquid with an average of only 24 V DC. When an uncontrolled arc forms at very high-voltages, arc flashes can produce deafening noises, supersonic concussive forces, super-heated shrapnel, temperatures far greater than the Sun's surface and intense, high-energy radiation capable of specialise nearby materials.

In summary, and in addition to the potential for electric shock, careless work on electrical systems (at any voltage) can result in

- fire
- explosion
- chemical release
- gases/fumes

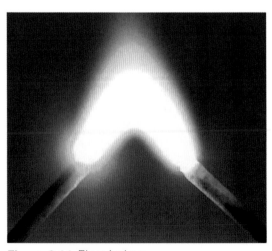

Figure 2.14 Electrical arc

2.2.8 **Protection devices**

The first-line protection against high-voltages includes direct methods such as

- enclosure (keeping things covered)
- insulation (always orange-coloured)
- location (positions to prevent accidental tampering)

The four main indirect methods to protect against high-voltages and excess current flow are

- fuses
- miniature circuit breaker (MCB)
- residual current device (RCD)
- residual current breaker with overcurrent (RCBO)

These four methods will now be outlined in more detail.

A fuse is a deliberate weak link in an electrical circuit that acts as a sacrificial device to provide overcurrent protection. It is a metal wire or strip that melts when too much current flows through it, therefore breaking the circuit. Short circuits, overloading, mismatched loads or device failure are the main reasons for excessive current.

Figure 2.15 Miniature blade fuse (actual size is about 15 mm)

A miniature circuit breaker (MCB) does the same job as a fuse in that it automatically switches off the electrical circuit during an overload condition. MCBs are more sensitive to overcurrent than fuses. They are quick and easy to reset by simply switching them back on. Most MCBs work by either the thermal or electromagnetic effect of overcurrent. The thermal operation is achieved with a bimetallic strip. The deflection of the bimetallic strip as it is heated by excess current releases a mechanical latch and opens the circuit. The electromagnetic type uses magnetism to operate the contacts. During short-circuit condition, the sudden increase in current causes a plunger to move and open the contacts.

> **Definition**
> MCB: miniature circuit breaker.

An residual current device (RCD) is designed to prevent fatal electric shock if a live connection is touched. RCDs offer a level of personal protection that ordinary fuses and circuit breakers cannot provide. If it detects electricity flowing down an unintended path, such as through a person who has touched a live part, the device will switch the circuit off very quickly, significantly reducing the risk of death or serious injury.

> **Definition**
> RCD: residual current device.

A residual current breaker with overcurrent (RCBO) is a type of circuit breaker designed to protect life in the same way as the RCD, but it also protects against an overload on a circuit. An RCBO will normally have two circuits for detecting an imbalance and an overload but use the same interrupt method.

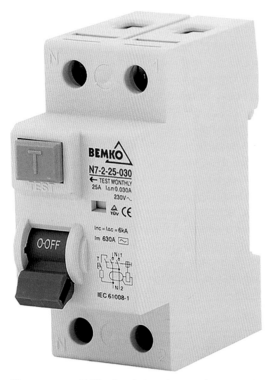

Figure 2.16 RCD circuit breaker – often used as part of a domestic charger

2.3 Safe work process

2.3.1 Risks of working with EVs

EVs introduce hazards into the workplace in addition to those normally associated with the repair and maintenance of vehicles, roadside recovery and other vehicle-related activities. These include

▶ the presence of high-voltage components and cabling capable of delivering a fatal electric shock
▶ the storage of electrical energy with the potential to cause explosion or fire
▶ components that may retain a dangerous voltage even when a vehicle is switched off
▶ electric motors or the vehicle itself that may move unexpectedly due to magnetic forces within the motors
▶ manual handling risks associated with battery replacement

▶ the potential for the release of explosive gases and harmful liquids if batteries are damaged or incorrectly modified
▶ the possibility of people being unaware of vehicles moving because, when electrically driven, they are silent in operation
▶ the potential for the electrical systems on the vehicle to affect medical devices such as pacemakers and insulin controllers

Control of substances hazardous to health (COSHH) regulations with regard to hazardous battery chemicals and compounds exist to assist with how to deal with leakage from battery packs.

However, batteries are in protective cases, and even if the case is damaged, batteries will not leak a significant amount of electrolyte. NiMH and Li-ion are dry-cell batteries and will only produce a few drops per cell if crushed. Some models may leak coolant, and this should not be confused with electrolyte.

2.3.2 Work categories

Four categories of work have been identified by the UK Health and Safety Executive (HSE). These are

▶ valeting, sales and other lower risk activities
▶ incident response, including emergency services and vehicle recovery
▶ maintenance and repair, excluding high-voltage electrical systems
▶ high-voltage electrical system repair and maintenance

Based on information from HSE, these categories are outlined next with the suggested primary controls.

2.3.3 Valeting, sales and other lower risk activities

Remote operation keys that only need to be close to the vehicle for the vehicle to be powered up should be kept away from vehicles. This is to prevent the vehicle from accidentally moving. People who move these vehicles around the workplace should be aware that others may not hear it approaching them. Similarly, people who work around Evs should be aware that they may move without warning. Pressure washing has the potential to damage high-voltage electrical components and cables. High-voltage cables are usually coloured orange. Refer to guidance from manufacturers before valeting any under body areas, including the engine bay.

2.3.4 Incident response, including emergency services and vehicle recovery

Vehicles should be visually checked for signs of damage to high-voltage electrical components or cabling (usually coloured orange). Consider whether the integrity of the battery is likely to have been compromised. Shorting or loss of coolant may present ignition sources in the event of fuel spillage. If the vehicle is damaged or faulty and if safe to do so, isolate the high-voltage battery system using the isolation device on the vehicle. Refer to the manufacturer's instructions for guidance. During any recovery onto a towing vehicle, the remote operation key should be removed to a suitable distance and the standard 12/24 V battery disconnected to prevent the vehicle from being activated/ started. Have access to reliable sources of information for specific vehicle types. For example, mobile data terminals used by fire and rescue services or by reference to manufacturer's data. Avoid towing EV vehicles unless it can be determined that it is safe to do so. Dangerous voltages can be generated by movement of the drive wheels.

2.3.5 Maintenance and repair, excluding high-voltage electrical systems

Refer to vehicle-specific sources of information from the manufacturer and trade bodies to identify precautions necessary to prevent danger. Remote operation keys should be

kept away from the vehicle to prevent any accidental operation of electrical systems and accidental movement of the vehicle. Keys should be locked away with access controlled by the person working on the vehicle. If the key is required during the work, the person working on the vehicle should check that the vehicle is in a safe condition before the key is retrieved. Visually check the vehicle for signs of damage to high-voltage cabling (usually coloured orange) or electrical components before starting any work on the vehicle. Unless a specific task requires the vehicle to be specialised, always isolate or disconnect the high-voltage battery in accordance with the manufacturer's instructions. Determine the locations of high-voltage cables before carrying out tasks such as panel replacement, cutting or welding. Take appropriate precautions to prevent them from being damaged.

2.3.6 Working on high-voltage electrical systems

Refer to vehicle-specific sources of information from the manufacturer and trade bodies to identify precautions you need to implement that are necessary to prevent danger. Remote operation keys should always be kept away from the vehicle to prevent any accidental operation of electrical systems and accidental movement of the vehicle. Keys should be locked away with access controlled by the person working on the vehicle. If the key is required during the work, the person working on the vehicle should check that the vehicle is in a safe condition before the key is retrieved. Visually check the vehicle for signs of damage to high-voltage electrical components or cabling (usually coloured orange). High-voltage systems should be isolated (that is, the power disconnected and secured such that it cannot be inadvertently switched back on) and proven dead by testing before any work is undertaken. Always isolate and lock off the source of electricity in accordance with manufacturer's instructions. You must always test and prove that any high-voltage cable or electrical component is dead before carrying out any work on it.

Even when isolated, vehicle batteries and other components may still contain large amounts of energy and retain a high-voltage. Only suitable tools and test equipment should be used. These may include electrically insulated tools and test equipment compliant with GS38.

Some electronic components may store dangerous amounts of electricity even when the vehicle is off and the battery isolated. Refer to manufacturer's data on how to discharge stored energy.

> ### Safety First
> Some electronic components may store dangerous amounts of electricity even when the vehicle is off and the battery isolated.

There may be circumstances (e.g., after collision damage) where it has not been possible to fully isolate the high-voltage electrical systems and to discharge the stored energy in the system. Refer to the manufacturer's instructions about what control measures should be implemented before attempting to carry out further remedial work.

Battery packs are susceptible to high temperatures. The vehicle will typically be labelled advising of its maximum temperature, and this should be considered when carrying operations such as painting, where booth temperatures may exceed this limit. Measures should be implemented to alleviate any potential risks (e.g., by removing the batteries or by providing insulation to limit any temperature increase in the batteries).

Working on live electrical equipment should *only* be considered when there is no other way for work to be undertaken. Even then, it should only be considered if it is both reasonable and safe to do so. You should consider the risks for working on this live equipment and

implement suitable precautions, including, as a final measure, the use of personal protective equipment (PPE). Refer to the manufacturer's instructions for precautions when working live, including the PPE requirements.

It may be necessary to locate the vehicle within an area that can be secured such that people who could be put at risk are not able to approach the vehicle. Warning signs should be used to make people aware of the dangers.

The following section will outline some further practical advice relating to this level of work.

2.3.7 Work processes

There are four stages or processes to consider for safe EV working:

- before
- during
- interrupted
- after

2.3.7.1 Before work starts

Electrical work should not start until protective measures have been taken against electric shock, short circuits and arcs. Work should not be performed on live parts of electrical systems and equipment. For this purpose, these systems and equipment must be placed in the non-live state prior to, and for the duration of, the work. This is achieved by following these three steps (but always check manufacturer's data):

1. Isolate
- Switch off the ignition.
- Remove service plug/maintenance connector or switch off main battery switch.
- Remove fuses/low-voltage battery where appropriate.

2. Safeguard against reconnection
- Remove the ignition key and prevent specialised access to it.
- Store the service plug/maintenance connector against specialised access/

safeguard the main battery switch against reconnection, for example, by means of a lock of some sort.

▶ Observe any additional manufacturer or company instructions.

3. Verify the non-live state

▶ The provisions of the vehicle manufacturer must be observed for verification of the non-live state.

▶ Suitable voltage testers or test apparatus specific to the manufacturer must be used.

▶ Until the non-live state has been verified, the system is to be assumed to be live.

▶ Wait an additional five minutes before performing any maintenance procedures on the system to allow any storage capacitors to be discharged.

▶ Make sure that the junction board terminal voltage is nearly 0 V.

Safety First

A non-live state is achieved by following three steps:

1. *Isolate.*
2. *Safeguard against reconnection.*
3. *Verify the non-live state.*

Figure 2.17 Work area fenced off and a warning sign in place

2.3.7.2 During work

During work, it is important to prevent shorts to earth and short circuits between components – even though they are disconnected. Remember a battery that has been disconnected is still live! If necessary, you should shroud or cover adjacent live parts. Always use suitable PPE and, when appropriate, use insulated tools when performing service procedures to the high-voltage system. This precaution will prevent accidental short circuits.

2.3.7.3 Interruption to work

When maintenance procedures have to be interrupted while some high-voltage components are uncovered or disassembled, make sure that the vehicle remains de-energised and isolated; the ignition remains turned off and the key is removed; and the battery module switch or service connector stays off. No untrained persons should have access to the area to prevent any unintended touching of the components.

2.3.7.4 Completion of work

Once the work has been completed, the safety process can be lifted. All tools, materials and other equipment must first be removed from the site of the work and the hazard area. Guards removed before the start of work must be properly replaced and warning signs removed. Before switching on the battery module or following the re-energisation process and after repairs have been completed, make sure that all terminals have been tightened to the specified torque and no high-voltage wires or terminals have been damaged. The insulation resistance between each high-voltage terminal of the part you disassembled and the vehicle's body should also have been checked.

29

Figure 2.18 Maintenance connector (green component)

Figure 2.19 Battery lift (Source: Snap-on/Sun Electric)

Safety First

Working on electric vehicles is not dangerous IF the previous guidelines and manufacturer's procedures are followed. Before starting work, check the latest information – DON'T take chances.

2.3.8 Lifting techniques

Lifting heavy components such as an EV drive motor is no different from lifting any other heavy item such as an engine on an ICE vehicle. Suitable cranes and straps should be used and manufacturer's recommendations followed at all times.

Battery lifting adds a new dimension, however, to our existing list. The process for supporting and removing a battery pack is relatively simple, but as always, manufacturer's guidance must be followed. Suitable equipment is essential as batteries can weigh in at around 450 kg and, in some cases, up to double this figure.

The Sun Electric and Hybrid Battery Lift (Figure 2.19) is a good example of a dedicated lifting table to assist with mounting and removing batteries from electric and hybrid vehicles. It features a dual-speed, electrohydraulic lifting system and has a total

lifting capacity of 1200 kg within a lifting range of 800 mm to 1810 mm.

Pins and pads are included for positioning and securing components (Figure 2.20). The platform extends to a maximum of 325 mm to accommodate large loads.

Stop, slow rise, standard rise and down functions are all available on hand-held lift controls. The suspend platform can tilt within a height of 40 mm. The lower platform can move laterally within a distance of 20 mm. There are

Figure 2.20 Battery lift in use with adaptors (Source: Snap-on/Sun Electric)

also reserved holes (100 mm x 100 mm) for positioning and mounting components with specialised fixtures.

2.3.9 Annual inspections

In most countries, Evs are subject to an annual safety check just like any other vehicle. Here in the UK, this is the ministry of transport (MOT) test. It is first carried out three years after purchasing a new vehicle, then annually. A recent consultation was carried out further to a suggestion that the first test should move to the fourth year. The theory being that EVs are less likely to go wrong – this was strongly opposed by anyone who had any understanding of the trade (over 75% of respondents)!

Extensive research on MOT data actually shows that an EV is more likely to fail its first test after three years when compared with an ICE vehicle. Tyres being the key area of failure.

Hybrid and electric vehicles do not pose any more of a risk than a conventional vehicle when carrying out an inspection such as an MOT test. However, all electrically powered vehicles pose a theoretical risk because of the high-voltage used in these vehicles. This is usually between 350 V and 600 V.

Unless stated otherwise, you do not need to fully isolate the high-voltage systems when carrying out an inspection. Nonetheless, you should avoid touching any high-voltage components and wiring when carrying out an inspection.

Many mild hybrids use 48 V systems. These are not high-voltage and may have blue wiring insulation instead of orange. You should still avoid touching the wiring on these systems.

The high-voltage components on hybrid and electric vehicles are usually inaccessible. When they are accessible, they will be well-insulated and do not present a high risk. However, it is important to visually check high-voltage

components and wiring as well as all the normal inspection points.

2.4 Hazard management

To manage hazards, you should also be able to identify vehicles and components and be aware of high-voltages as covered in other sections of this book.

2.4.1 Initial assessment

First responders should carry out an initial visual risk assessment. Personal protection should be worn. Steps should then be taken to secure the safety of themselves and others at incident scenes involving EVs. For example, people who may be at risk are

▶ occupants
▶ on-lookers
▶ recovery personnel
▶ emergency service personnel

Safety First

If you call for help, make sure you tell the operator that an EV is involved.

Vehicles damaged by fire or impact can result in these risks:

▶ electric shock
▶ burns
▶ arc flash
▶ arc blast
▶ fire
▶ explosion
▶ chemicals
▶ gases/fumes

It may, therefore, be necessary to implement evacuation procedures and site protection.

Safety First

First responders should carry out an initial visual risk assessment.

Figure 2.21 Battery fire

2.4.2 **Fire**

There are substantial differences in the designs of EVs and their component parts from different manufacturers. Having information specific to the manufacturer and the vehicle being worked on is important in identifying what actions are necessary to work safely.

As well as the obvious need to take personal precautions, incorrect maintenance operations when dealing with EV high-voltage systems can result in damage to the vehicle, other people and property.

When working on EVs, normal protection should be used such as wing covers, floor mats and other items. Disposal of waste materials is no different from ICE vehicles, with the exception of the high-voltage battery. If high-voltage battery stacks/modules develop a fault, it is possible that thermal runaway can occur. Thermal runaway refers to a situation where an increase in temperature changes the conditions in a way that causes a further increase in temperature, often leading to a destructive result. It is a kind of uncontrolled positive feedback. Fires may occur in an EV high-voltage battery, or a fire may extend to the battery. Most EV batteries currently on the road are Li-ion, but NiMH batteries are popular too. There is a range of guidance concerning the tactics for dealing with EVs in which the battery is burning. However, the general consensus is that the use of water or other standard agents does not present an electrical hazard to firefighting personnel.

If a high-voltage battery catches fire, it will require a large, sustained volume of water. If a Li-ion high-voltage battery is involved in fire, there is a possibility that it could reignite after extinguishment, so thermal imaging should be used to monitor the battery. If there is no immediate threat to life or property, it is recommended that a battery fire be allowed to burn out.

Another further consideration with an EV fire is that the automatic built-in measures to prevent electrocution from high-voltages may be compromised. For example, the normally open relays for the high-voltage system could possibly fail in a closed position if they sustain damage due to heat.

> **Safety First**
> If a battery fire occurs, it should be left to the fire brigade to deal with.

2.5 **Tools and equipment**

2.5.1 **Introduction**

By way of an introduction, Tables 2.3 and 2.4 lists some of the basic words and descriptions relating to tools and equipment.

2.5.2 **Hand tools**

Using hand tools is something you will learn by experience, but an important first step is to understand the purpose of the common types. This section, therefore, starts by listing some of the more popular tools, with examples of their use, and ends with some general advice and instructions. Practise until you understand the use and purpose of the following tools when working on vehicles.

General advice and instructions for the use of hand tools (supplied by Snap-on):

▶ Only use a tool for its intended purpose.
▶ Always use the correct size tool for the job you are doing.

Table 2.3 Equipment terminology

Hand tools equipment terminology	Spanners and hammers and screwdrivers and all the other basic bits! Description
Accuracy	Careful and exact, free from mistakes or errors and adhering closely to a standard.
Calibration	Checking the accuracy of a measuring instrument.
Code reader or scanner	This device reads the 'black and white balls' mentioned before or the on-off electrical signals and converts them into language we can understand.
Combined diagnostic and information system	Usually now PC-based, these systems can be used to carry out tests on vehicle systems, and they also contain an electronic workshop manual. Test sequences guided by the computer can also be carried out.
Dedicated test equipment	Some equipment will only test one specific type of system. The large manufacturers supply equipment dedicated to their vehicles. For example, a diagnostic device that plugs in to a certain type of fuel injection engine control unit.
Oscilloscope	The main part of the 'scope' is the display, which is like a TV or computer screen. A scope is a voltmeter, but instead of readings in numbers, it shows the voltage levels by a trace or mark on the screen. The marks on the screen can move and change very quickly, allowing us to see the way voltages change.
Serial port	A connection to an electronic control unit, a diagnostic tester or computer for example. Serial means the information is passed in a 'digital' string, like pushing black and white balls through a pipe in a certain order.
Special tools	A collective term for items not held as part of a normal tool kit. Or items required for just one specific job.
Test equipment	In general, this means measuring equipment. Most tests involve measuring something and comparing the result of that measurement with data. The devices can range from a simple ruler to an engine analyser.

Figure 2.22 Snap-on tool kit

▶ Pull a spanner or wrench rather than pushing whenever possible.
▶ Do not use a file, or similar, without a handle.
▶ Keep all tools clean and replace them in a suitable box or cabinet.
▶ Do not use a screwdriver as a pry bar.
▶ Look after your tools and they will look after you!

2.5.3 Test equipment

To remove, refit and adjust components to ensure the vehicle system operates within specification is a summary of almost all the work you will be doing. The use, care, calibration and storage of test equipment are, therefore, very important. In this sense, 'test equipment' means

▶ measuring equipment – such as a micrometer
▶ hand instruments – such as a spring balance
▶ electrical meters – such as a digital multimeter or an oscilloscope

The operation and care of this equipment will vary with different types. I suggest, therefore, that you should always read the manufacturer's instructions carefully before use or if you have a problem.

The following list sets out good general guidelines:

▶ Follow the manufacturer's instructions at all times.

Table 2.4 Hand tools

Hand tool	Example uses and/or notes
Adjustable spanner (wrench)	An ideal standby tool, useful for holding one end of a nut and bolt.
Air wrench	These are often referred to as wheel guns. Air-driven tools are great for speeding up your work, but it is easy to damage components because an air wrench is very powerful. Only special, extra strong, high-quality sockets should be used.
Blade (engineer's) screwdriver	Simple common screw heads. Use the correct size!
Hammer	Anybody can hit something with a hammer, but exactly how hard and where is a great skill to learn!
Hexagon socket spanner	Sockets are ideal for many jobs where a spanner can't be used. In many cases, a socket is quicker and easier than a spanner. Extensions and swivel joints are also available to help reach that awkward bolt.
Levers	Used to apply a very large force to a small area. If you remember this, you will realise how, if incorrectly applied, it is easy to damage a component.
Open-ended spanner	Use for nuts and bolts where access is limited or a ring spanner can't be used.
Pliers	These are used for gripping and pulling or bending. They are available in a wide variety of sizes. These range from snipe nose, for electrical work, to engineer's pliers for larger jobs such as fitting split pins.
Pozidrive, Phillips and crosshead screwdrivers	Better grip is possible, particularly with the Pozidrive, but learn not to confuse the two very similar types. The wrong type will slip and damage will occur.
Ring spanner	The best tool for holding hexagon bolts or nuts. If fitted correctly, it will not slip and damage both you and the bolt head.
Socket wrench	Often contains a ratchet to make operation far easier.
Special purpose wrenches	Many different types are available. As an example, mole grips are very useful tools because they hold like pliers but can lock in position.
Torque wrench	Essential for correct tightening of fixings. The wrench can be set in most cases to 'click' when the required torque has been reached. Many fitters think it is clever not to use a torque wrench. Good technicians realise the benefits.
Torx®	Similar to a hexagon tool like an Allen key but with further flutes cut in the side. It can transmit good torque.

Figure 2.23 Digital multimeter in use

▶ Handle with care: Do not drop it and keep the instrument in its box.
▶ Ensure regular calibration: Check for accuracy.
▶ Understand how to interpret results: If in doubt, ask!

My favourite piece of test equipment is the PicoScope. This is an oscilloscope that works through a computer. It will test all engine management systems and other electrical and electronic devices. Check out www.picoauto.com for more information. Figure 2.24 shows a signal from an inductive sensor taken using the PicoScope.

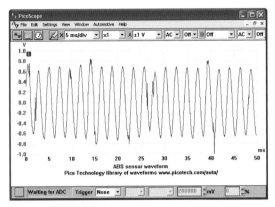

Figure 2.24 Waveform display from the PicoScope (Source: Pico Technology Ltd)

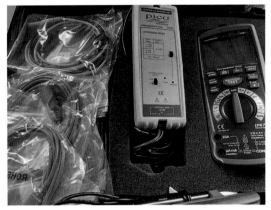

Figure 2.26 Electric vehicle PicoScope kit, cables, multimeter, and differential probe

Key Fact

An oscilloscope draws a graph of voltage against time.

For EV work, electrical meters such as a voltmeter should be rated to a minimum 1000 V.

There is a range of different multimeters available and choosing the right one is essential. Indeed, like me, you may choose to own more than one, and, of course, you get what you pay for.

Figure 2.25 Electric vehicle PicoScope kit main unit and amp clamps

The Fluke 78 automotive meter shown in Figure 2.23 is several years old now but perfectly functional and ideal for low-voltage applications (included here because it has served me well – newer CAT III/IV versions are available!). It has the following features and specifications:

▶ volts, amps, continuity and resistance
▶ frequency for pulsed-DC and AC frequency tests
▶ duty cycle to verify operation of sensors and actuator supply signals
▶ direct reading of dwell for 3-, 4-, 5-, 6-, and 8-cylinder engines
▶ temperature readings up to 999 °C (1,830 °F) using the thermocouple bead probe and adapter plug
▶ min/max recording that works with all meter functions
▶ precision analog bar graph
▶ RPM inductive pickup for both conventional and distributorless (DIS) ignitions
▶ 10 MΩ input impedance
▶ CAT II 300 V (ideal for low-voltage automotive and even mains voltages but not recommended for H/EV high-voltage use)

Meters and their leads have category ratings that give the voltage levels up to which they

35

Figure 2.27 Automotive meter accessories (test leads, 600 A clamp, plug lead RPM, temperature thermocouple)

Figure 2.28 CAT IV meter in use (on low-voltage)

are safe to use. Category (CAT) ratings can be a little confusing, but there is one simple rule of thumb: Select a multimeter rated to the highest category in which it could possibly be

used. In other words, err on the side of safety. Table 2.6 lists some of the different ratings.

The voltages listed in Table 2.6 are those that the meter will withstand without damage or risk to the user. A test procedure (known as IEC 1010) is used and takes three main criteria into account:

▶ steady-state working voltage
▶ peak-impulse transient voltage
▶ source impedance

These three criteria together will tell you a multimeter's true *voltage withstand* values. However, this is confusing because it can look as if some 600 V meters offer more protection than 1000 V ones.

> **Definition**
> Impedance is the total opposition to current flow in an AC circuit (in a DC circuit it is described as resistance).

Within a category, a higher working voltage is always associated with a higher transient voltage. For example, a CAT III 600 V meter is tested with 6000 V transients, while a CAT III 1000 V meter is tested with 8000 V transients. This indicates that they are different, and that the second meter clearly has a higher rating. However, the 6000 V transient CAT III 600 V meter and the 6000 V transient CAT II 1000 V meter are not the same even though the transient voltages are. This is because the source impedance has to be considered.

Table 2.5 CAT ratings

Category	Working voltage (voltage withstand)		Peak impulse (transient voltage withstand)		Test source impedance
CAT I	600 V		2500 V		30 Ω
CAT I	1000 V		4000 V		30 Ω
CAT II	600 V		4000 V		12 Ω
CAT II	1000 V		6000 V		12 Ω
CAT III	600 V		6000 V		2 Ω
CAT III	1000 V		8000 V		2 Ω
CAT IV	600 V		8000 V		2 Ω

Figure 2.29 CAT III 1000 V and CAT IV 600 V meter

Figure 2.30 CAT III 1000 V and CAT IV 600 V leads

To measure voltage, the meter is connected in parallel with the circuit. The most common measurement on a vehicle is DC voltage. Remember to set the range of the meter (some are auto-ranging) and, if in doubt, start with a higher range and work downwards.

To measure resistance, the meter must be connected across (in parallel with) the component or circuit under test. However, the circuit must be switched off or isolated. If not, the meter will be damaged. Likewise, because an ohmmeter causes a current to flow, there are some circuits, such as Hall effect sensors, that can be damaged by the meter.

Ohm's Law (I = V/R) shows that the 2 Ω test source for CAT III will have six times the current of the 12 Ω test source for CAT II. The CAT III 600 V meter, therefore, offers better transient protection compared to the CAT II 1000 V meter, even though, in this case, the voltage rating appears to be lower.

The combination of working voltage and category determines the total 'voltage withstand' rating of a multimeter (or any other test instrument), including the very important 'transient voltage withstand' rating. Remember, for working on vehicle high-voltage systems, you should choose a *CAT III or CAT IV meter AND leads*.

There are lots of different options or settings available when using a multimeter, but the three most common measurements are: voltage (volts), resistance (ohms) and current (amps).

Figure 2.31 Checking a simple resistor

Figure 2.32 Inductive ammeter clamp on a high-voltage cable (measuring the current drawn by the EV cabin heater)

Current can be measured in two ways:

1. Connect the meter in series with the circuit (in other words, break the circuit and reconnect it through the meter).
2. Use an inductive amp clamp around the wire (Figure 2.32), which is a safer way to measure but is less accurate at low values.

This internal resistance of a meter can affect the reading it gives on some circuits. It is recommended that this should be a minimum of 10 MΩ, which ensures accuracy because the meter only draws a very tiny (almost insignificant) current. This stops the meter loading the circuit and giving an inaccurate reading, and it prevents damage to sensitive circuits in an electronic control unit (ECU), for example.

However, the very tiny current draw of a good multimeter can also be a problem. A supply voltage of say 12 V can be shown on a meter when testing a circuit but does not prove the integrity of the supply. This is because a meter with a 10 MΩ internal resistance connected to a 12 V supply will only cause a current of 1.2 µA (I=V/R) – that's 1.2 millionths of an amp, which will not cause any noticeable voltage loss even if there is an unwanted resistance of several thousand ohms in the supply circuit. A test lamp can be connected in parallel with the meter to load the circuit (make more current flow) but should be used carefully

so you don't damage sensitive electronic switching circuits that may be present.

Voltmeters can display a 'ghost' voltage rather than zero when the leads are open circuit.

In other words, if checking the voltage at an earth/chassis connection, we would expect a 0 V reading. However, the meter will also display zero before it is connected, so how do we know the reading is correct when it is connected?

> **Key Fact**
>
> Voltmeters can display a 'ghost' voltage rather than zero when the leads are open circuit.

The answer is to shake the multimeter leads, a 'ghost' voltage will fluctuate (Figure 2.33), a real voltage will not!

An insulation tester does exactly as its name suggests. On automotive systems, this test is mostly used on electric and hybrid vehicles. Refer to manufacturer's information before carrying out any tests on the high-voltage system – and be TechSafe™.

The device shown in Figure 2.34 is known as a Megger. It is a multimeter but also able to use up to 1000 V to test the resistance of insulation on a wire or component. A reading well in excess of 10 MΩ (Figure 2.34) is what we would normally expect if the insulation is in

Figure 2.33 Ghost voltage caused by shaking the red lead

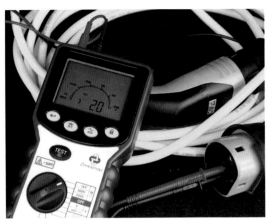

Figure 2.34 Checking insulation resistance between conductors in an EV charging lead (in this case the reading is >20 GΩ)

good order. The high-voltage is used because it puts the insulation under pressure and will show up faults that would not be apparent if you used an ordinary ohmmeter.

Take care when using insulation testers. The high-voltage used for the test will not kill a fit person because it cannot sustain a significant current flow, but it still hurts!

2.5.4 Workshop equipment

Safety First

For EV work, electrical meters such as a voltmeter should be rated to a minimum CAT III or CAT IV.

In addition to hand tools and test equipment, most workshops will also have a range of equipment for lifting and supporting as well as electrical or air-operated tools. Table 2.7 lists some examples of common workshop equipment together with typical uses.

Table 2.6 Examples of workshop equipment

Equipment	Common use
Ramp or hoist	Used for raising a vehicle off the floor. They can be a two-post wheel-free type, and other designs include four-post and scissor types where the mechanism is built into the workshop floor.
Jack and axle stands	A trolley jack is used for raising part of a vehicle such as the front or one corner or side. It should always be positioned under suitable jacking points, axle or suspension mountings. When raised, stands must always be used in case the seals in the jack fail causing the vehicle to drop.
Impact Gun	A high-pressure air supply is common in most workshops, but the air gun has mostly been replaced by a battery impact gun for removing wheel nuts or bolts. Note that when replacing wheel fixings, it is essential to use a torque wrench.
Electric drill	The electric drill is just one example of electric power tools used for automotive repair. Note that it should never be used in wet or damp conditions.
Parts washer	There are a number of companies that supply a parts washer and change the fluid it contains at regular intervals.
Steam cleaner	Steam cleaners can be used to remove protective wax from new vehicles as well as to clean grease, oil and road deposits from cars in use. They are supplied with electricity, water and a fuel to run a heater, so caution is necessary.
Electric welder	There are a number of forms of welding used in repair shops. The most common is metal inert gas (MIG).
Gas welder	Gas welders were once popular, but more often induction heat tools are used instead, for example, when heating a flywheel ring gear.
Engine crane	A crane of some type is essential for removing the engine on most vehicles. It usually consists of two legs with wheels that go under the front of the car and a jib that is operated by a hydraulic ram. Chains or straps are used to connect to or wrap around the engine.
Transmission jack	On many vehicles the transmission is removed from underneath. The car is supported on a lift and then the transmission jack is rolled underneath.

Figure 2.35 Trolley jack and axle stands (Source: Snap-on Tools)

1000 V or, in some cases, up to 10,000 V in compliance with EN 60900. In fact, the tools were tested individually at 10,000 V for 10 seconds at a time.

The range comprises a full complement of insulated tools, including ratchets, sockets, screwdrivers, spanners, T-wrenches, pliers and an insulated torque wrench. Latex insulation gloves, protective outer gloves and a secure roller cabinet are also available.

An important safety feature of the EV tool range is the two-step colour-code system.

If any of the orange-coloured outer insulation material is missing, a bright yellow interior is exposed, clearly indicating to the technician that the tool is no longer safe for use.

2.5.5 High-voltage tools

Many manufacturers have designed ranges of tools that are designed to protect mechanics from the high-voltage systems in electric vehicles. There are now several companies that produce high-quality insulated tools and associated equipment that protects up to

2.5.6 Scanner

On-board diagnostics (OBD) is a generic term referring to a vehicle's self-diagnostic and reporting system. OBD systems give the vehicle owner or a technician access to information for various vehicle systems.

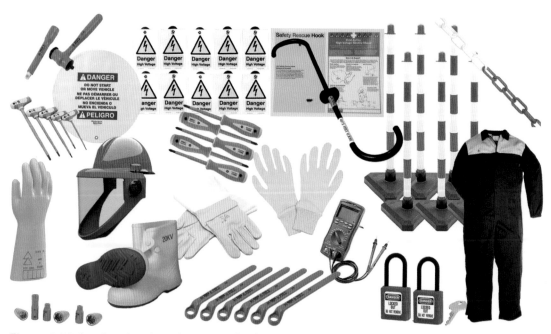

Figure 2.36 Insulated tools and a range of safety equipment (Source: EINTAC, https://eintac.com)

Definition

OBD: on-board diagnostics.

Diagnostic trouble codes (DTCs), or fault codes, are stored by an on-board computer diagnostic system. These codes are stored when, for example, a sensor in the car produces a reading that is outside its expected range.

DTCs identify a specific problem area and are a guide as to where a fault might be occurring within the vehicle. Parts or components should not be replaced with reference only to a DTC. No matter what some customers may think, the computer does not tell us exactly what is wrong! For example, if a DTC reports a sensor fault, replacement of the sensor is unlikely to resolve the underlying problem. The fault is more likely to be caused by the systems that the sensor is monitoring but can also be caused by the wiring to the sensor.

DTCs may also be triggered by faults earlier in the operating process. For example, a dirty MAF sensor might cause the car to overcompensate its fuel-trim adjustments. As a result, an oxygen sensor fault may be set as well as a MAF sensor code. All but the most basic DIY scanners will translate the codes and present the text so there's no need to refer to a list.

In addition to OBD codes, most scanners now communicate with all the electronic modules on the data bus and display appropriate DTCs and live data.

A useful feature has been added to a number of scanners, which is the ability to show the controller area network (CAN) topology. Other scanners are available, but I used a TopDon Phoenix Pro to get the results shown in Figure 2.39. The vehicle was a Volkswagen Golf GTE (2018). Figure 2.40 shows the same thing but with the fault codes cleared. The colour-coded topology display shows all vehicle systems and their status on one page. Note

Figure 2.37 Diagnostic data link connector (DLC)

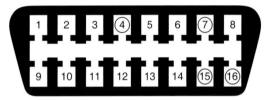

Figure 2.38 Connector pin-out: 4, battery ground/earth; 7, K line; 15, L line; t, battery positive

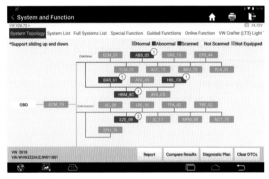

Figure 2.39 System topology showing possible DTCs on some of the systems (Source: www.diagnosticconnections.com)

how it also shows the relationship between the different nodes and the different networks:

▶ CAN-Drive
▶ CAN-Comfort
▶ CAN-Info

The display shows the structure of communication between modules, and you

41

Figure 2.40 System topology showing no DTCs after the faults have been rectified or cleared (Source: www.diagnosticconnections. com)

can then press on a system module to display any DTCs and access the system.

The benefit of this topology display is that as well as showing the relationship between different networks, you can also see which modules are on which network so that, for example, several errors on one network may indicate a common cause. It is also very convenient to tap on a node where an error is shown and open full details of that part.

Some of the abbreviations used in Figures 2.39 and 2.40 are industry standard or at least in common use. Others are proprietary to Volkswagen. Pressing the item opens a screen with more details so it is usually easy to work out what the item is.

2.6 Equality, diversity and inclusion (EDI)

2.6.1 Authors' comment

EDI stands for equity, diversity and inclusion. It's a movement to change the landscape, giving everyone the opportunities and support they need. Tom and Hayley are proud to support EDI and have signed the IMI pledge: https://tide.theimi.org.uk/ equity-diversity-and-inclusion/sign-the-pledge

Please consider adding your name to the list of people who have pledged to upholding the values of EDI and are committed to bringing a positive change to our industry.

2.6.2 Institute of the Motor Industry (IMI)

At the time of writing, the automotive sector currently has the highest vacancy rate in over 20 years, and the rate is climbing – There are now 25,000 vacancies in the sector. The IMI carried out extensive research, which highlighted the need for a closer look at how diverse and inclusive the automotive industry is so we can work to tackle the turbulent times that the sector is facing.

With its Diversity Task Force partners, it created a one-stop shop for all things EDI-related within the automotive sector. Using this, you can quickly educate and update your knowledge with the wide range of resources, tools, and guidance so you can make a positive change in your workplace. Let's all work together to tackle the challenges underrepresented groups face.

Follow this link to learn more about the thought-provoking work the IMI have carried out, along with some of the leading employers within the sector: https://tide.theimi.org.uk/ equity-diversity-and-inclusion

2.6.3 What is equality?

At its core, equality means fairness: We must ensure that individuals, or groups of individuals, are not treated less favourably because of their protected characteristics. Equality also means equality of opportunity: We must also ensure that those who may be disadvantaged can get the tools they need to access the same, fair opportunities as their peers.

2.6.4 What is diversity?

Diversity is recognising, respecting and celebrating each other's differences. A diverse environment is one with a wide range of backgrounds and mindsets, which allows

Figure 2.41 Diversity is not only the right thing to do, but it hugely benefits our businesses

for an empowered culture of creativity and innovation.

2.6.5 **What is inclusion?**

Inclusion means creating an environment where everyone feels welcome and valued. An inclusive environment can only be created once we are more aware of our unconscious biases and have learned how to manage them.

2.6.6 **What are the protected characteristics?**

The following are the legal protected characteristics, under *The Equality Act* 2010:

▶ age
▶ disability
▶ gender reassignment
▶ marriage and civil partnership
▶ pregnancy and maternity
▶ race
▶ religion or belief
▶ sex
▶ sexual orientation

Discrimination on the grounds of any of these characteristics is illegal. Discrimination can take many forms including direct discrimination, indirect discrimination, bullying, harassment and victimisation.

CHAPTER 3

Electrical and electronic principles

3.1 Basic electrical principles

3.1.1 Introduction

To understand electricity properly, we must start by finding out what it really is. This means we must think very small. The molecule is the smallest part of something that can be recognised as that particular matter. Subdivision of the molecule results in atoms, which are the smallest part of matter. An element is a substance that comprises atoms of one kind only.

The atom consists of a central nucleus made up of protons and neutrons. Around this nucleus orbit electrons, like planets around a sun. The neutron is a very small part of the nucleus. It has equal positive and negative charges and is therefore neutral and has no polarity. The proton is another small part of the nucleus and is positively charged. The neutron is neutral and the proton is positively charged, which means that the nucleus of the atom is positively charged. The electron is an even smaller part of the atom and is negatively charged. It orbits the nucleus and is held in orbit by the attraction of the positively charged proton. All electrons are similar no matter what type of atom they come from.

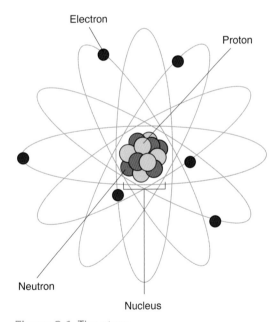

Figure 3.1 The atom

When atoms are in a balanced state, the number of electrons orbiting the nucleus equals the number of protons. The atoms of some materials have electrons that are easily detached from the parent atom and can, therefore, join an adjacent atom. In so doing, these atoms move an electron from the parent atom to another atom (like polarities repel)

DOI: 10.1201/9781003431732-3

Figure 3.2 Electronic components have made technology such as this 950 V Audi Formula-e car possible (Source: Audi Media)

and so on through material. This is a random movement, and the electrons involved are called free electrons.

Materials are called conductors if the electrons can move easily. In some materials, it is extremely difficult to move electrons from their parent atoms. These materials are called insulators.

> **Definition**
> Materials are called conductors if the electrons can move easily.

> **Definition**
> Materials are called insulators if the electrons are difficult to move.

3.1.2 Electron and conventional flow

If an electrical pressure (electromotive force or voltage) is applied to a conductor, a directional movement of electrons will take place (for example, when connecting a battery to a wire). This is because the electrons are attracted

to the positive side and repelled from the negative side. Certain conditions are necessary to cause an electron flow:

▶ a pressure source (e.g., from a battery or generator)
▶ a complete conducting path in which the electrons can move (e.g., wires).

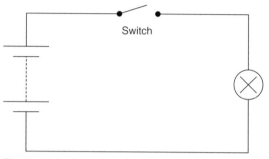

Figure 3.3 A simple electrical circuit

An electron flow is termed an electric current. Figure 3.3 shows a simple electric circuit where the battery positive terminal is connected, through a switch and lamp, to the battery negative terminal. With the switch open, the chemical energy of the battery will remove electrons from the positive terminal to the negative terminal via the battery. This leaves the positive terminal with fewer electrons and the negative terminal with a surplus of electrons. An electrical pressure, therefore, exists between the battery terminals.

> **Definition**
> An electron flow is termed an electric current.

With the switch closed, the surplus electrons at the negative terminal will flow through the lamp back to the electron-deficient positive terminal. The lamp will light, and the chemical energy of the battery will keep the electrons moving in this circuit from negative to positive. This movement from negative to positive is called the electron flow and will continue while the battery supplies the pressure – in other words, while it remains charged.

▶ Electron flow is from negative to positive.

It was once thought, however, that current flowed from positive to negative, and this convention is still followed for most practical purposes. Therefore, although this current flow is not correct, the most important point is that we all follow the same convention.

▶ Conventional current flow is said to be from positive to negative.

Key Fact

Conventional current flow is said to be from positive to negative.

3.1.3 Effects of current flow

When a current flows in a circuit, it can produce only three effects:

▶ heating
▶ magnetic
▶ chemical

The heating effect is the basis of electrical components such as lights and heater plugs. The magnetic effect is the basis of relays and motors and generators. The chemical effect is the basis for electroplating and battery charging.

In the circuit shown in Figure 3.5, the chemical energy of the battery is first converted to electrical energy, then into heat energy in the lamp filament.

The three electrical effects are reversible. Heat applied to a thermocouple will cause a small electromotive force and, therefore, a small current to flow. Practical use of this is mainly in instruments. A coil of wire rotated in the field of a magnet will produce an electromotive force and can cause current to flow. This is the basis of a generator. Chemical action, such as in a battery, produces an electromotive force, which can cause current to flow.

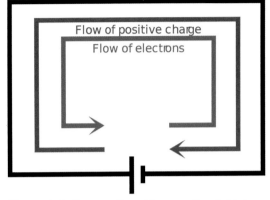

Figure 3.4 Current flow (Source: Romtobbi)

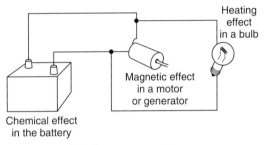

Figure 3.5 A bulb, motor and battery – heating, magnetic and chemical effects

Key Fact

The three electrical effects are reversible.

3.1.4 Fundamental quantities

In Figure 3.5, the number of electrons through the lamp every second is described as the rate of flow. The cause of the electron flow is the electrical pressure. The lamp produces an opposition to the rate of flow set up by the electrical pressure. Power is the rate of doing work, or changing energy from one form to another. These quantities, as well as several others, are given names as shown in Table 3.1.

If the voltage pressure applied to the circuit was increased but the lamp resistance stayed the same, then the current would also increase. If the voltage was maintained constant but the lamp was changed for

3 Electrical and electronic principles

Table 3.1 Quantities, symbols and units

Name	Definition	Symbol	Common formula	Unit name	Abbreviation
Electrical charge	One coulomb is the quantity of electricity conveyed by a current of one ampere in one second.	Q	$Q = It$	Coulomb	C
Electrical flow or current	The number of electrons past a fixed point in one second.	I	$I = V/R$	Ampere	A
Electrical pressure	A pressure of 1 volt applied to a circuit will produce a current flow of 1 amp if the circuit resistance is 1 ohm.	V	$V = IR$	Volt	V
Electrical resistance	This is the opposition to current flow in a material or circuit when a voltage is applied across it.	R	$R = V/I$	Ohm	Ω
Electrical conductance	Ability of a material to carry an electrical current. One Siemens equals 1 amp per volt. It was formerly called the mho or reciprocal ohm.	G	$G = 1/R$	Siemens	S
Current density	The current per unit area. This is useful for calculating the required conductor cross-sectional areas.	J	$J = I/A$ (A = area)		A m^2
Resistivity	A measure of the ability of a material to resist the flow of an electric current. It is numerically equal to the resistance of a sample of unit length and unit cross-sectional area, and its unit is the ohm metre. A good conductor has a low resistivity ($1.7 \times 10^8\,\Omega$ m copper); an insulator has a high resistivity ($10^{15}\,\Omega$ m polyethane).	ρ (rho)	$R = \rho L/A$ (L = length; A = area)	Ohm metre	Ω m
Conductivity	The reciprocal of resistivity.	σ (sigma)	$\sigma = 1/\rho$	Ohm^1 metre^1	Ωs^1 m^1
Electrical power	When a voltage of 1 volt causes a current of 1 amp to flow, the power developed is 1 watt.	P	$P = IV$ $P = I^2R$ $P = V^2/R$	Watt	W
Capacitance	Property of a capacitor that determines how much charge can be stored in it for a given potential difference between its terminals.	C	$C = Q/V$ $C = \varepsilon A/d$ (A = plate area, d = distance between, ε = permittivity of dielectric)	Farad	F

Table 3.1 (Continued)

Name	Definition	Symbol	Common formula	Unit name	Abbreviation
Inductance	Where a changing current in a circuit builds up a magnetic field, which induces an electromotive force either in the same circuit and opposing the current (self-inductance) or in another circuit (mutual inductance).	L	$i = \dfrac{V}{R}\left(1 - e^{-Rt/L}\right)$ (i = instantaneous current, R = resistance, L = inductance, t = time, e = base of natural logs)	Henry	H
Magnetic field strength or intensity	Magnetic field strength is one of two ways that the intensity of a magnetic field can be expressed. A distinction is made between magnetic field strength H and magnetic flux density B.	H	$H = B/\mu_0$ (μ_0 being the magnetic permeability of space)	Amperes per metre	A/m (An older unit for magnetic field strength is the oersted: 1 A/m = 0.01257 oersted)
Magnetic flux	A measure of the strength of a magnetic field over a given area.	Φ (phi)	$\Phi = \mu H A$ (μ = magnetic permeability, H = magnetic field intensity, A = area)	Weber	Wb
Magnetic flux density	The density of magnetic flux, 1 tesla is equal to 1 weber per square metre. Also measured in Newton-metres per ampere (Nm/A).	B	$B = H/A$ $B = H \times \mu$ (μ = magnetic permeability of the substance, A = area)	Tesla	T

one with a higher resistance, the current would decrease. Ohm's Law describes this relationship.

Ohm's Law states that in a closed circuit 'current is proportional to the voltage and inversely proportional to the resistance'. When 1 V causes 1 A to flow, the power used (P) is 1 W.

Using symbols this means:

$$\text{Voltage} = \text{Current} \times \text{Resistance}$$
$$(V = IR) \text{ or } (R = V/I) \text{ or } (I = V/R)$$

$$\text{Power} = \text{Voltage} \times \text{Current}$$
$$(P = VI) \text{ or } (I = P/V) \text{ or } (V = P/I)$$

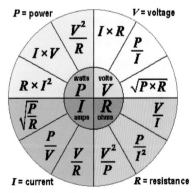

Figure 3.6 Formula wheel

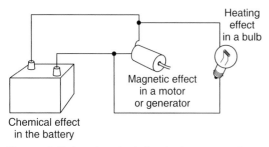

Heating effect in a bulb

Magnetic effect in a motor or generator

Chemical effect in the battery

Figure 3.7 An electrical circuit demonstrating links between voltage, current, resistance and power

3.1.5 **Describing electrical circuits**

Three descriptive terms are useful when discussing electrical circuits.

▶ open circuit: This means the circuit is broken, therefore, no current can flow.
▶ short circuit: This means that a fault has caused a wire to touch another conductor and the current uses this as an easier way to complete the circuit.
▶ high resistance: This means a part of the circuit has developed a high resistance (such as a dirty connection), which will reduce the amount of current that can flow.

3.1.6 **Conductors, insulators and semiconductors**

All metals are conductors. Gold, silver, copper and aluminium are among the best and are frequently used. Liquids that will conduct an electric current are called electrolytes. Insulators are generally nonmetallic and include rubber, porcelain, glass, plastics, cotton, silk, wax paper and some liquids.

Some materials can act as either insulators or conductors, depending on conditions. These are called semiconductors and are used to make transistors and diodes.

> **Key Fact**
> Gold, silver, copper, and aluminium are among the best conductors.

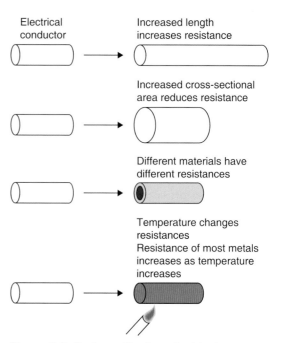

Electrical conductor

Increased length increases resistance

Increased cross-sectional area reduces resistance

Different materials have different resistances

Temperature changes resistances
Resistance of most metals increases as temperature increases

Figure 3.8 Factors affecting electrical resistance

3.1.7 **Factors affecting the resistance of a conductor**

In an insulator, a large voltage applied will produce a very small electron movement. In a conductor, a small voltage applied will produce a large electron flow or current. The amount of resistance offered by the conductor is determined by a number of factors.

▶ length: the greater the length of a conductor, the greater is the resistance
▶ cross-sectional area: the larger the cross-sectional area, the smaller the resistance
▶ material: the resistance offered by a conductor will vary according to the material from which it is made, which is known as the resistivity or specific resistance of the material
▶ temperature: most metals increase in resistance as temperature increases

3.1.8 Resistors and circuit networks

Good conductors are used to carry the current with minimum voltage loss due to their low resistance. Resistors are used to control the current flow in a circuit or to set voltage levels. They are made of materials that have a high resistance. Resistors intended to carry low currents are often made of carbon. Resistors for high currents are usually wire wound.

> **Key Fact**
>
> Resistors are used to control the current flow in a circuit or to set voltage levels.

Resistors are often shown as part of basic electrical circuits to explain the principles involved. The circuits shown as Figure 3.9 are equivalent. In other words, the circuit just showing resistors is used to represent the other circuit.

When resistors (or bulbs in this case) are connected so that there is only one path (Figure 3.10) for the same current to flow through each bulb, they are connected in series and the following rules apply:

▶ Current is the same in all parts of the circuit.

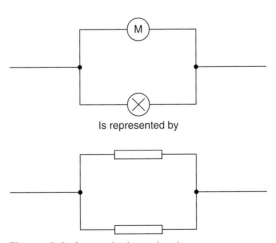

Is represented by

Figure 3.9 An equivalent circuit

Figure 3.10 Series circuit

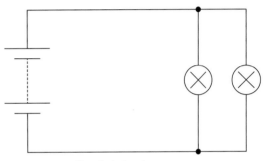

Figure 3.11 Parallel circuit

▶ The applied voltage equals the sum of the volt drops around the circuit.
▶ Total resistance of the circuit (RT) equals the sum of the individual resistance values ($R_1 + R_2$ etc.).

When resistors or bulbs are connected such that they provide more than one path (Figure 3.11 shows two paths) for the current to flow through and have the same voltage across each component, they are connected in parallel and the following rules apply.

▶ The voltage across all components of a parallel circuit is the same.
▶ The total current equals the sum of the current flowing in each branch.
▶ The current splits up depending on each component resistance.
▶ The total resistance of the circuit (RT) can be calculated by:

$$1/R_T = 1/R_1 + 1/R_2 \text{ or}$$
$$R_T = (R_1 \times R_2) / (R_1 + R_2)$$

51

3.1.9 Magnetism and electromagnetism

Magnetism can be created by a permanent magnet or by an electromagnet (remember: it is one of the three effects of electricity). The space around a magnet in which the magnetic effect can be detected is called the magnetic field. The shape of magnetic fields in diagrams is represented by flux lines or lines of force.

Some rules about magnetism:

▶ Unlike poles attract. Like poles repel.
▶ Lines of force in the same direction repel sideways; in the opposite direction, they attract.
▶ Current flowing in a conductor will set up a magnetic field around the conductor. The strength of the magnetic field is determined by how much current is flowing.
▶ If a conductor is wound into a coil or solenoid, the resulting magnetism is the same as that of a permanent bar magnet.

Electromagnets are used in motors, relays and fuel injectors, to name just a few applications. Force on a current-carrying conductor in a magnetic field is caused because of two magnetic fields interacting. This is the basic principle of how a motor works. Figure 3.12 shows a representation of these magnetic fields.

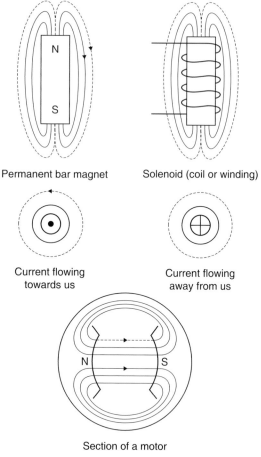

Permanent bar magnet Solenoid (coil or winding)

Current flowing towards us Current flowing away from us

Section of a motor

Figure 3.12 Magnetic fields

Key Fact

Force on a current-carrying conductor in a magnetic field is caused because of two magnetic fields interacting.

3.1.10 Electromagnetic induction

Basic laws of electromagnetic induction:

▶ When a conductor cuts or is cut by magnetism, a voltage is induced in the conductor.
▶ The direction of the induced voltage depends on the direction of the magnetic field and the direction in which the field moves relative to the conductor.
▶ The voltage level is proportional to the rate at which the conductor cuts or is cut by the magnetism.

This effect of induction, meaning that voltage is made in the wire, is the basic principle of how generators such as the alternator on a car work. A generator is a machine that converts mechanical energy into electrical energy. Figure 3.13 shows a wire moving in a magnetic field.

Definition

A generator is a machine that converts mechanical energy into electrical energy.

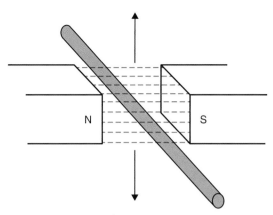

Figure 3.13 Induction

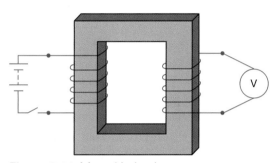

Figure 3.14 Mutual induction

3.1.11 **Mutual induction**

If two coils (known as the primary and secondary) are wound onto the same iron core, then any change in magnetism of one coil will induce a voltage into the other. This happens when a current to the primary coil is switched on and off. If the number of turns of wire on the secondary coil is more than the primary, a higher voltage can be produced. If the number of turns of wire on the secondary coil is fewer than the primary, a lower voltage is obtained. This is called 'transformer action' and is the principle of the ignition coil. Figure 3.14 shows the principle of mutual induction.

The value of this 'mutually induced' voltage depends on

▶ the primary current
▶ the turns ratio between primary and secondary coils
▶ the speed at which the magnetism changes

> **Key Fact**
> Transformer action is the principle of the ignition coil. It is also used in a DC–DC converter.

3.1.12 **Basic motor principle**

Electric motors operate because of the interaction between magnetic fields. The magnetic fields can be created by windings or be permanent magnets, but one of them must be electromagnetic. The force from the two fields is made to act in such a way as to cause a shaft to turn (see Figure 3.15). Electric motors can be powered by DC, such as from batteries, or by AC, such as the mains power grid or inverters.

A generator and a motor are effectively the same machine except generators convert mechanical energy into electrical energy, and motors convert electrical energy into mechanical energy.

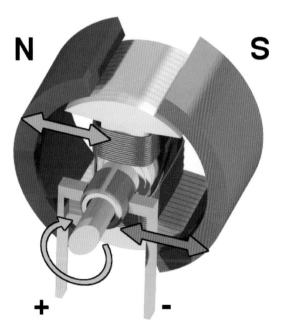

Figure 3.15 Basic electric motor principle (Source: Wikimedia)

53

3.2 Definitions and laws

3.2.1 Ohm's law

▶ For most conductors, the current that will flow through them is directly proportional to the voltage applied to them.

The ratio of voltage to current is referred to as resistance. If this ratio remains constant over a wide range of voltages, the material is said to be 'ohmic'.

> **Key Fact**
> The ratio of voltage to current is referred to as resistance.

$$R = V/I$$

where I = current in amps, V = voltage in volts, R = resistance in ohms.

Georg Simon Ohm was a German physicist well-known for his work on electrical currents.

3.2.2 Lenz's law

▶ The EMF induced in an electric circuit always acts in a direction so that the current it creates around the circuit will oppose the change in magnetic flux which caused it.

Lenz's law gives the direction of the induced EMF resulting from electromagnetic induction. The 'opposing EMF' is often described as a 'back EMF'.

The law is named after the Estonian physicist Heinrich Lenz.

3.2.3 Kirchhoff's laws

Kirchhoff's 1st law:

▶ The current flowing into a junction in a circuit must equal the current flowing out of the junction.

This law is a direct result of the conservation of charge; no charge can be lost in the junction, so any charge that flows in must also flow out.

Kirchhoff's 2nd law:

▶ For any closed-loop path around a circuit, the sum of the voltage gains and drops always equals zero.

This is effectively the same as the series circuit statement that the sum of all the voltage drops will always equal the supply voltage.

Gustav Robert Kirchhoff was a German physicist; he also discovered cesium and rubidium.

3.2.4 Faraday's law

▶ Any change in the magnetic field around a coil of wire will cause an EMF (voltage) to be induced in the coil.

It is important to note here that no matter how the change is produced, the voltage will be generated. In other words, the change could be produced by changing the magnetic field strength, moving the magnetic field towards or away from the coil, moving the coil in or out of the magnetic field, rotating the coil relative to the magnetic field and so on!

Michael Faraday was a British physicist and chemist, well-known for his discoveries of

Figure 3.16 The original Faraday induction coil in the museum at the Royal Institution in London

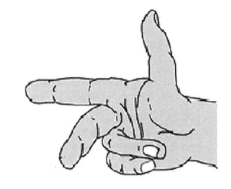

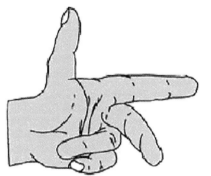

Figure 3.17 Fleming's rules

electromagnetic induction and of the laws of electrolysis.

3.2.5 Fleming's rules

▶ In an electrical machine, the first finger lines up with the magnetic Field, the second finger lines up with the Current and the thumb lines up with the Motion.

Fleming's rules relate to the direction of the magnetic field, current and motion in electrical machines. The left hand is used for motors and the right hand for generators (remember: gener-righters).

John Fleming was an English electrical engineer and physicist.

3.2.6 Ampere's law

▶ For any closed-loop path, the sum of the length elements × the magnetic field in the direction of the elements = the permeability × the electric current enclosed in the loop.

In other words, the magnetic field around an electric current is proportional to the electric current which creates it, and the electric field is proportional to the charge which creates it.

André Marie Ampère was a French scientist known for his significant contributions to the study of electrodynamics.

3.2.7 Summary

It was tempting to conclude this section by stating some of Murphy's laws, for example:

▶ If anything can go wrong, it will go wrong.
▶ You will always find something in the last place you look.
▶ In a traffic jam, the lane on the motorway that you are not in always goes faster.

. . . but I decided against it!

> **Definition**
> Murphy's law: If anything can go wrong, it will go wrong.

3.3 Electronic components

3.3.1 Introduction

This section, describing the principles and applications of various electronic circuits, is not intended to explain their detailed operation. The intention is to describe briefly how the circuits work and, more importantly, how and where they may be utilised in vehicle applications.

The circuits described are examples of those used, and many pure electronics books are available for further details. Overall, an understanding of basic electronic principles will help to show how electronic control units work, ranging from a simple interior light delay unit to the most complicated engine management system.

3.3.2 **Components**

The main devices described here are often known as discrete components. Figure 3.18 shows the symbols used for constructing the circuits shown later in this section. A simple and brief description follows for many of the components.

Resistors are probably the most widely used component in electronic circuits. Two factors must be considered when choosing a suitable resistor, namely the ohms value and the power rating. Resistors are used to limit current flow and provide fixed voltage drops. Most resistors used in electronic circuits are made from small carbon rods, and the size of the rod determines the resistance. Carbon resistors have a negative temperature coefficient (NTC) and this must be considered for some applications. Thin film resistors have more stable temperature properties and are constructed by depositing a layer of carbon onto an insulated former such as glass. The resistance value can be manufactured very accurately by spiral grooves cut into the carbon film. For higher power applications, resistors are usually wire wound. This can, however, introduce inductance into a circuit. Variable forms of most resistors are available in either linear or logarithmic forms. The resistance of a circuit is its opposition to current flow.

> **Definition**
> Negative temperature coefficient (NTC): as temperature increases, resistance decreases.

A capacitor is a device for storing an electric charge. In its simple form, it consists of two plates separated by an insulating material. One plate can have excess electrons compared with the other. On vehicles, its main uses are for reducing arcing across contacts and for radio interference suppression circuits as well as in electronic control units. Capacitors are described as two plates separated by a dielectric. The area of the plates A, the distance between them d and the permittivity (ε) of the dielectric determine the value of capacitance. This is modelled by the equation: Metal foil sheets insulated by a type of paper are often used to construct capacitors. The sheets are rolled up together inside a tin can. To achieve higher values of capacitance, it is necessary to reduce the distance between the plates in order to keep the overall size of the device manageable. This is achieved by immersing one plate in an electrolyte to deposit a layer of oxide typically 100 μm thick, thus ensuring a higher capacitance value. The problem, however, is that this now makes the device polarity-conscious and only able to withstand low voltages. Variable capacitors are available that are varied by changing either of the variables given in the previous equation. The unit of capacitance is the farad (F). A circuit has a capacitance of one farad (1 F) when the charge stored is one coulomb and the potential difference is 1 V. Figure 3.19 shows a capacitor charged up from a battery.

Diodes are often described as one-way valves, and, for most applications, this is an acceptable description. A diode is a simple PN junction allowing electron flow from the N-type material (negatively biased) to the P-type material (positively biased). The materials are usually constructed from doped silicon. Diodes are not perfect devices, and a voltage of about 0.6 V is required to switch the diode on in its forward-biased direction. Zener diodes are very similar in operation, with the exception that they are designed to break down and conduct in the reverse direction at a predetermined voltage. They can be thought of as a type of pressure-relief valve.

> **Definition**
> Diodes are often described as one-way valves.

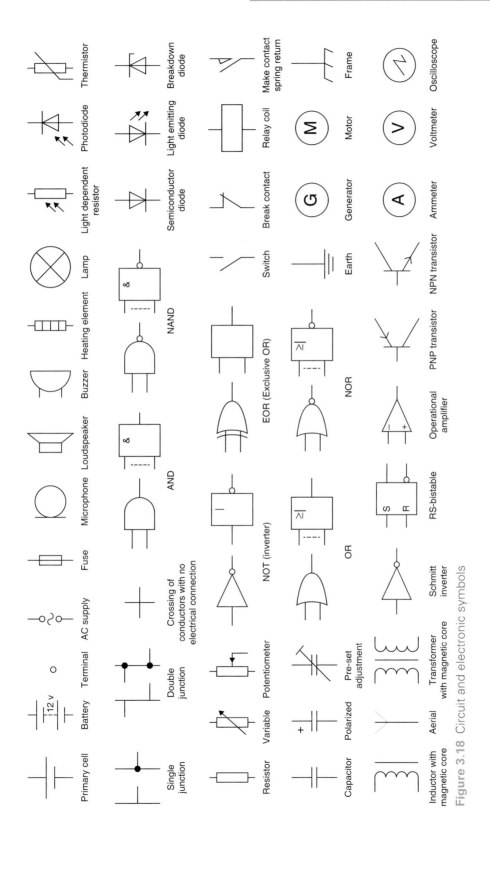

Figure 3.18 Circuit and electronic symbols

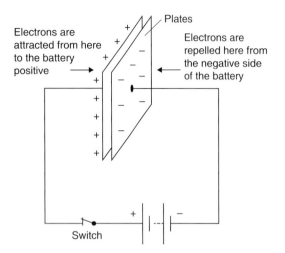

Electrons are attracted from here to the battery positive →

Plates

Electrons are repelled here from the negative side of the battery ←

Switch

When the switch is opened, the plates stay as shown. This is simply called 'charged up'

Figure 3.19 A capacitor charged up

Figure 3.20 IGBT packages

Transistors are the devices that have allowed the development of today's complex and small electronic systems. They replaced the thermal-type valves. The transistor is used as either a solid-state switch or as an amplifier. Transistors are constructed from the same P- and N-type semiconductor materials as the diodes and can be either made in NPN or PNP format. The three terminals are known as the base, collector and emitter. When the base is supplied with the correct bias, the circuit between the collector and emitter will conduct. The base current can be of the order of 200 times less than the emitter current. The ratio of the current flowing through the base compared with the current through the emitter (I_e/I_b) is an indication of the amplification factor of the device and is often given the symbol β (beta).

Another type of transistor is the field effect transistor (FET). This device has higher input impedance than the bipolar type described above. FETs are constructed in their basic form as N-channel or P-channel devices. The three terminals are known as the gate, source and drain. The voltage on the gate terminal controls the conductance of the circuit between the drain and the source.

A further and important development in transistor technology is the insulated gate bipolar transistor (IGBT). The insulated gate bipolar transistor (Figure 3.20) is a three-terminal power semiconductor device, noted for high efficiency and fast switching. It switches electric power in many modern appliances: electric cars, trains, variable speed refrigerators, air conditioners and even stereo systems with switching amplifiers. Since it is designed to rapidly turn on and off, amplifiers that use it often synthesise complex waveforms with pulse-width modulation and low-pass filters.

Definition

IGBT: insulated gate bipolar transistor.

Inductors are most often used as part of an oscillator or amplifier circuit. In these applications, it is essential for the inductor to be stable and to be of reasonable size. The basic construction of an inductor is a coil of wire wound on a former. It is the magnetic effect of the changes in current flow that gives this device the properties of inductance. Inductance is a difficult property to control, particularly as the inductance value increases due to magnetic coupling with other devices. Enclosing the coil in a can will reduce this, but eddy currents are then induced in the can and this affects the overall inductance value. Iron cores are used to increase the inductance value as this changes the permeability of the core. However, this also allows for adjustable devices by moving the position of the core. This only allows the value to change by a few percent but is useful for tuning a circuit. Inductors, particularly of higher values, are often known as chokes and may be used in DC circuits to smooth the voltage. The value of inductance is the henry (H). A circuit has an inductance of one henry (1 H) when a current, which is changing at one ampere per second, induces an electromotive force of one volt in it.

Power metal oxide semiconductor field effect transitors (MOSFET) are also used as switching devices in inverters. This is because they are extremely fast, which means lower switching losses. Only a small gate current is needed to operate a MOSFET. These devices are, therefore, often considered to be the ideal switch. They can be found in most power supplies, DC–DC converters and low-voltage motor controllers. Super-junction MOSFETs are also now being used to increase the efficiency and reduce the size of power supplies even further.

Definition

MOSFET: metal oxide semiconductor field effect transistor.

Ceramic substrates can play a significant role in MOSFETs and IGBTs by boosting performance, minimising power losses and controlling temperature. These advanced substrates are designed to carry higher

currents, provide higher voltage isolation and operate over a wide temperature range. Ceramic substrates are used for IGBTs and MOSFETs.

3.3.3 Integrated circuits

Integrated circuits are constructed on a single slice of silicon often known as a substrate. In an integrated circuit, some of the components mentioned previously can be combined to carry out various tasks such as switching, amplifying and logic functions. In fact, the components required for these circuits can be made directly on the slice of silicon. The great advantage of this is not just the size of the integrated circuits but the speed at which they can be made to work due to the short distances between components. Switching speeds in excess of 1 MHz are typical.

There are four main stages in the construction of an integrated circuit. The first of these is oxidisation by exposing the silicon slice to an oxygen stream at a high temperature. The oxide formed is an excellent insulator.

The next process is photoetching, where part of the oxide is removed. The silicon slice is covered in a material called a photoresist, which, when exposed to light, becomes hard. It is now possible to imprint the oxidised silicon slice, which is covered with photoresist, by a pattern from a photographic transparency. The slice can now be washed in acid to etch back to the silicon those areas that were not protected by being exposed to light. The next stage is diffusion, where the slice is heated in an atmosphere of an impurity such as boron or phosphorus, which causes the exposed areas to become N- or P-type silicon. The final stage is epitaxy, which is the name given to crystal growth. New layers of silicon can be grown and doped to become N- or P-type as before. It is possible to form resistors in a similar way and small values of capacitance can be achieved. It is not possible to form any useful inductance on a chip. Figure 3.21 shows the construction and Figure 3.22

Figure 3.21 Integrated circuit components

Figure 3.24 Integrated circuits

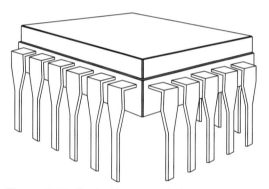

Figure 3.22 Typical integrated circuit package

shows a representation of the 'packages' that integrated circuits are supplied in for use in electronic circuits.

The range and types of integrated circuits now available are so extensive that a chip is available for almost any application. The integration level of chips has now reached and, in many cases, is exceeding that of very large scale integration (VLSI). This means there can be more than 100,000 active elements on one chip. Development in this area is moving so fast that often the science of electronics is now concerned mostly with choosing the correct combination of chips, and discreet components are only used as final switching or power output stages.

> **Key Fact**
>
> Today's microprocessors have many millions of gates and billions of individual transistors (well in excess of VLSI).

Figure 3.23 Discrete components

CHAPTER 4

Electric vehicle technology

4.1 Electric vehicle layouts

4.1.1 Identifying electric vehicles

There are several types of electric vehicles, but many look very similar to their non-electric counterparts, so look out for the badging! The following pictures show some common types:

Figure 4.2 Plug-in hybrid car – VW Golf GTE (Source: Volkswagen)

Figure 4.1 Hybrid car – Toyota Prius (Source: Toyota Media)

Figure 4.3 Pure electric car – Nissan LEAF (Source: Nissan)

DOI: 10.1201/9781003431732-4

Figure 4.4 Concept electric motorcycle (Source: Harley Davidson Media)

Figure 4.5 Commercial electric truck (Source: Volvo Media)

Figure 4.6 Passenger bus on charge

Figure 4.7 shows the general layout in block diagram form of an electric vehicle (EV). Note that because the drive batteries are a few hundred volts, a lower 12/24 V system is still required for 'normal' lighting and other systems.

4.1.2 Single motor

The 'classic' pure-EV layout is to use a single motor driving either the front or rear wheels. Most EVs of this type do not have a transmission gearbox because the motor operates at suitable torque throughout the speed range of the vehicle.

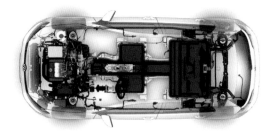

Figure 4.8 VW Golf-e layout with the motor at the front and the battery at the rear (Source: Volkswagen Media)

Figure 4.9 shows a sectioned view of a drive motor and the basic driveline consisting of a fixed ratio gear-set, the differential and driveshaft flanges.

Hybrid cars vary in layout, and this is examined in detail later in this chapter. However, the basic design is similar to the pure electric car mentioned before. The obvious difference being the addition of an ICE.

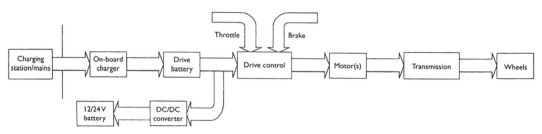

Figure 4.7 Generic electric vehicle layout

Figure 4.9 EV motor (Source: Volkswagen Media)

Figure 4.10 PHEV layout (Source: Volkswagen Media)

The motor for the plug-in hybrid is shown here where it forms part of the gearbox assembly. Motors used on light hybrids are sometimes described as integrated motor assist (IMA) because they form part of the flywheel.

This type of motor is shown as Figures 4.11 and 4.12.

Figure 4.11 PHEV engine, motor and gearbox (Source: Volkswagen Media)

Figure 4.12 Motor integrated with the engine flywheel (Source: Bosch Media)

4.1.3 **Wheel motors**

Protean Electric is a major player in the development of in-wheel EV motors. Their ProteanDrive consists of a permanent-magnet synchronous motor and integrated electronics (Figure 4.13). The electronics precisely control motor current, so each in-wheel motor can deliver the torque required in about a millisecond. In-wheel motors allow for

Figure 4.13 Mounting the motors in the wheels makes more space available for batteries, passengers or cargo (Source: Protean Electric)

torque-vectoring, which means different torques can be applied to different wheels. This can significantly improve handling.

The electronic circuitry fits within the overall motor package and shares cooling with the motor. The motor windings can have up to 90 A flowing through them. The heat that is developed, along with heat from the electronics, is controlled by coolant flowing through a channel in the motor housing. The coolant is in thermal contact with the electronic components and the motor windings. Those windings are encapsulated in epoxy resin, which also helps to conduct heat. Integration of the motor and drive electronics means a small motor can still generate significant power.

The challenge with wheel motors has always been to keep unsprung mass to a minimum. This improves ride quality for the driver and passengers and makes it easier for the suspension to keep the tyres in contact with the road. A vehicle powered by in-wheel electric motors will have significantly greater unsprung mass because the weight of a motor will be carried in each powered wheel. However, this can be all but mitigated by careful suspension design. Every technology on a car is a compromise of some sort. It is important to remember that while wheel motors have some disadvantages, they also have major advantages.

> **Key Fact**
>
> The challenge with wheel motors has always been to keep unsprung mass to a minimum.

The motor (Figure 4.14) is configured with the rotor on the outside. This arrangement places the gap between the stator and the rotor (across which the magnetic forces are developed) at the maximum radius available, thereby creating as much torque as possible within the confines of the wheel rim. This approach lets the motor

Figure 4.14 Protean Electric's in-wheel motor system is arguably simpler than a conventional electric automobile, which has constant-velocity joints, drive shafts and a centrally mounted transmission and motor (Source: Protean Electric)

develop sufficient torque without the need for gearing, which would have lowered efficiency and generated noise.

4.2 Hybrid electric vehicle layouts

4.2.1 Introduction

Hybrid vehicles use at least one electric drive motor in addition to the internal combustion engine (ICE). There are several different ways in which this can be combined and a number of different motors and engines. Note that for clarity we will generally refer to the ICE as an engine and the electric drive motor as a motor. Take care though because in other parts of the world, the ICE can be referred to as a motor!

There are three main objectives in the design of a hybrid vehicle:

1. Reduction in fuel consumption (and CO_2)
2. Reduction in emissions
3. Increased torque and power

A hybrid vehicle needs a battery to supply the motor; this is sometimes called an accumulator. The most common types are nickel–metal hydride (Ni–MH) or lithium-ion (Li-ion) and usually work at voltages between 200 V and 400 V.

The motors are generally permanent-magnet synchronous types and work in conjunction with an inverter (converts DC to AC, but more on this later). The key benefit of an electric drive is high

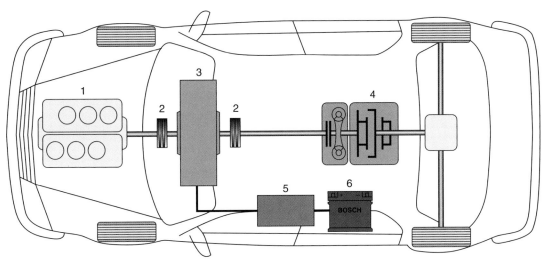

Figure 4.15 Hybrid layout (parallel): 1) ICE, 2) clutch, 3) motor, 4) transmission, 5) inverter, 6) battery

torque at low speed, so it is an ideal supplement to an ICE where the torque is produced at higher speeds. The combination, therefore, offers good performance at all speeds. The following graph shows typical results – note also that the engine capacity is reduced in the hybrid, but the result is still an improvement.

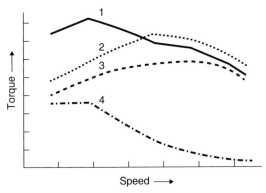

Figure 4.16 Comparing torque curves: 1) hybrid, 2) standard engine (1600 cc), 3) downsized engine (1200 cc), 4) motor (15 kW)

The result of the hybridisation of a motor and an engine is that it can always be operated (with suitable electronic control) at its optimum speed for reducing emissions and consumption while still producing good

torque. A smaller capacity engine can also be used (downsizing) in conjunction with a higher geared transmission so the engine runs at lower speed (downspeeding) but performance is maintained.

During braking, the motor becomes a generator, and the energy that would normally be wasted as heat from the brakes is converted into electrical energy and stored in the battery. This is used at a later stage, and, in some cases, the vehicle can run on electric only with zero emissions. Plug-in hybrids take this option even further.

Key Fact

On all types of EV, during braking, the motor becomes a generator and the energy that would normally be wasted as heat from the brakes is converted into electrical energy and stored in the battery – regenerative braking.

4.2.2 Classifications

Hybrids can be classified in different ways. There have been several different variations of this list,

65

Table 4.1 Hybrid functions

Classification/ function	Stop/start	Regeneration	Electrical assistance	Electric-only driving	Charging from a power socket
Stop/start system	√	√			
Mild hybrid	√	√	√		
Strong hybrid	√	√	√	√	
Plug-in hybrid	√	√	√	√	√

but the accepted classification is now that the vehicle fits in one of these four categories:

▶ stop/start system
▶ mild hybrid
▶ strong hybrid
▶ plug-in hybrid

The functions available from the different types are summarised in Table 4.1.

A stop/start system has the functions of stop/start as well as some regeneration. The control of the normal vehicle alternator is adapted to achieve this. During normal driving, the alternator operates with low output. During overrun, the alternator output is increased in order to increase the braking effect to increase power generation. Stopping the engine when idling saves fuel and reduces emissions. An uprated starter motor is needed to cope with the increased use as the vehicle is auto-started as the driver presses the accelerator. Fuel savings in the standard tests can be up to 5%.

The mild hybrid is like the previous system but also provides some assistance during acceleration, particularly at low speeds. Pure electric operation is not possible; the motor can propel the vehicle, but the engine is always running. Fuel savings in the standard tests can be up to 15%.

A strong hybrid takes all of the aforementioned functions further, and over short distances the engine can be switched off to allow pure electric operation. Fuel savings in the standard tests can be up to 30%.

The plug-in hybrid is a strong hybrid but with a larger high-voltage battery that can be charged from a suitable electrical power supply. Fuel

Figure 4.17 BMW 3-series plug-in hybrid

savings in the standard tests can be up to 70%, but note, this is over short distances.

4.2.3 **Operation**

In addition to a stop/start function and full electric operation, there are five main operating modes that a hybrid vehicle will use:

▶ startup
▶ acceleration
▶ cruising
▶ deceleration
▶ idle

These main modes and conditions are outlined in Figure 4.18. Further details of what is taking place during the different operating modes are outlined Figure 4.19. The operating modes are explained in even more detail in Table 4.2.

These descriptions relate generally to a light-hybrid, sometimes described as integrated motor assist (IMA). This is a parallel configuration discussed further in the next section. The technique used by most hybrid

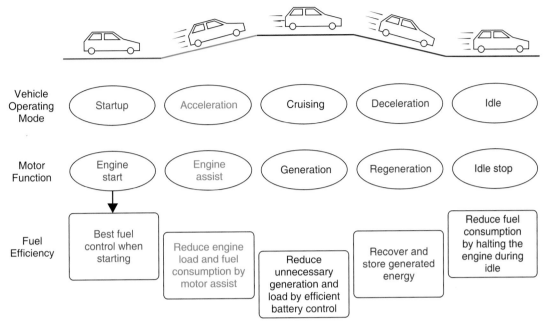

Figure 4.18 Hybrid vehicles operating modes

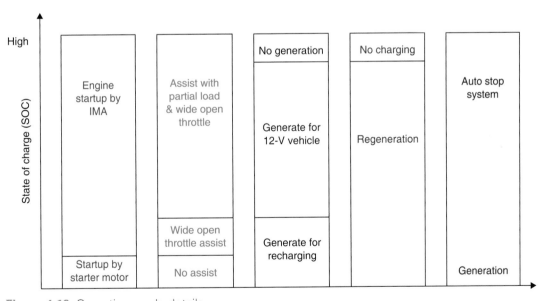

Figure 4.19 Operating mode details

cars can be thought of as a kinetic energy recovery system (KERS). This is because, instead of wasting heat energy from the brakes as the vehicle is slowed down, a large proportion is converted to electrical energy and stored in the battery as chemical energy. This is then used to drive the wheels, reducing the use of chemical energy from the fuel.

67

Table 4.2 Hybrid operating modes

Mode	Details
Startup	Under normal conditions, the motor will immediately start the engine at a speed of 1000 rpm. When the state of charge (SOC) of the high-voltage battery module is too low, when the temperature is too low or if there is a failure of the motor system, the engine will be cranked by the normal 12 V starter.
Acceleration	During acceleration, current from the battery module is converted to AC by an inverter and supplied to the motor. The motor output is used to supplement the engine output so that power available for acceleration is maximised. Current from the battery module is also converted to 12 V DC for supply to the vehicle electrical system. This reduces the load that would have been caused by a normal alternator and so improves acceleration. When the remaining battery state of charge is too low, but not at the minimum level, assist will only be available during wide-open throttle (WOT) acceleration. When the battery state of charge is reduced to the minimum level, no assist will be provided.
Cruising	When the vehicle is cruising and the battery requires charging, the engine drives the motor – which now acts as a generator. The resulting output current is used to charge the battery and is converted to supply the vehicle electrical system. When the vehicle is cruising and the high-voltage battery is sufficiently charged, the engine drives the motor. The generated current is converted to 12 V DC and only used to supply the vehicle electrical system.
Deceleration	During deceleration (during fuel cut), the motor is driven by the wheels such that regeneration takes place. The generated output is used to charge the high-voltage battery. Some vehicles cut the ICE completely. During braking (brake switch on), a higher amount of regeneration will be allowed. This will increase the deceleration force so the driver will automatically adjust the force on the brake pedal. In this mode, more charge is sent to the battery module. If the ABS system is controlling the locking of the wheels, an 'ABS-busy' signal is sent to the motor control module. This will immediately stop generation to prevent interference with the ABS system.
Idling	During idling, the flow of energy is similar to that for cruising. If the state of charge of the battery module is very low, the motor control system will signal the engine control module (ECM) to raise the idle speed to approximately 1100 rpm. On a stronger hybrid, the engine hardly ever idles as the motor will be used to move the vehicle and start the engine if necessary. Other vehicle functions such as AC can be run from the high-voltage battery if enough power is available.

Figure 4.20 Hybrid vehicle (Source: Audi Media)

4.2.4 **Configurations**

A hybrid power system for an automobile can have a series, parallel or power split configuration. With a series system, an engine drives a generator, which, in turn, powers a motor. The motor propels the vehicle. With a parallel system, the engine and motor can both be used to propel the vehicle. Most hybrids in current use employ a parallel system. The power split has additional advantages but is also more complex.

Hybrid electric vehicles are often described as being in categories P0 to P4, PS or EE as shown in Figure 4.21.

Key Fact

A hybrid power system for an automobile can have a series, parallel or power split configuration.

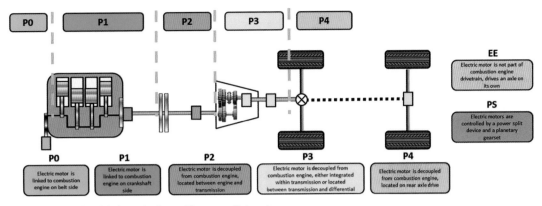

Figure 4.21 Hybrid descriptions (Source: Solvay)

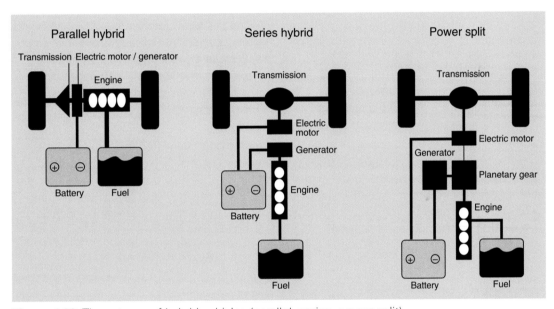

Figure 4.22 Three types of hybrid vehicles (parallel, series, power split)

There are three basic categories (Figure 4.22) but numerous configurations as manufacturers have developed different systems and ideas. However, it is now generally accepted that HEVs fall into one of the following descriptions:

▶ parallel hybrid with one clutch
▶ parallel hybrid with two clutches
▶ parallel hybrid with double-clutch transmission
▶ axle-split parallel hybrid
▶ series hybrid
▶ series-parallel hybrid
▶ power-split hybrid

The 'parallel hybrid with one clutch' is shown in Figure 4.23. This layout is a mild hybrid where the engine and motor can be used independently of each other, but the power flows are in parallel and can be added together to get the total drive power. The engine will run all the time the vehicle is driving, at the same speed as the motor.

The main advantage of this configuration is that the conventional drivetrain can be maintained. In most cases, only one motor is used and fewer adaptations are needed when converting a conventional system. However,

69

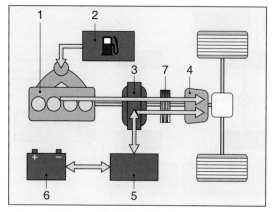

Figure 4.23 Parallel hybrid with one clutch (P1-HEV): 1) engine, 2) fuel tank, 3) motor (integrated motor generator – IMG, 4) transmission, 5) inverter, 6) battery, 7) clutch

Electronic control systems are used to determine when the clutches are operated; for example, the engine can be decoupled during deceleration to increase regenerative braking. It even allows the vehicle to go into 'sailing' mode, where it is slowed down only by rolling friction and aerodynamic drag.

If the engine-clutch is operated in such a way as to maintain torque, then the engine can be stopped and started using the clutch – a sophisticated bump start! Sensors and intelligent controls are needed to achieve this. In some cases, a separate starter motor is used, which can normally be dispensed with.

Adding the extra clutch in the previous system increases the length of the transmission, and this may be a problem, particularly in front wheel drive (FWD) cars. If a double-clutch transmission is used in the configuration shown in Figure 4.25, then this problem is overcome.

The motor is connected to a subunit of the transmission instead of the engine crankshaft or flywheel. These transmissions are also described as direct-shift gearboxes, or DSG. Pure electric driving is possible by opening the appropriate transmission clutch, or both

because the engine can be decoupled, it produces drag on overrun and reduces the amount of regeneration. Pure electric driving is not possible.

A parallel hybrid with two clutches is a strong hybrid and is an extension of the mild hybrid outlined before, except that the additional clutch allows the engine to be disconnected. This means pure electric use is possible.

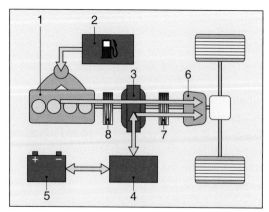

Figure 4.24 Parallel hybrid with two clutches (P2-HEV): 1) engine, 2) fuel tank, 3) motor (IMG), 4) inverter, 5) battery, 6) transmission, 7) clutch one, 8) clutch two

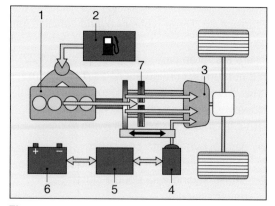

Figure 4.25 Parallel hybrid with double-clutch transmission: 1) engine, 2) fuel tank, 3) transmission, 4) motor, 5) inverter, 6) battery, 7) clutches

engine and motor can drive in parallel. The gear ratio between engine and motor can also be controlled in this system, allowing designers even greater freedom. Sophisticated electronic control, sensor and actuators are necessary.

Definition
DSG: direct shift gearbox.

The axle-split parallel hybrid is also a parallel drive even though the motor and engine are completely separated. As the name suggests, they drive an axle each. A semiautomatic transmission together with a stop/start system is needed with this layout. As the engine can be completely decoupled, this configuration is suitable for operation as a strong hybrid. It can effectively deliver all-wheel drive when the battery is charged, and, in some cases, to ensure this, an additional generator is fitted to the engine to charge the high-voltage battery even when the vehicle is stationary.

A series hybrid configuration is a layout where the engine drives a generator (alternator) that charges the battery that powers a motor that drives the wheels! A series configuration is always a strong hybrid since all the previously stated functions are possible (Table 4.1).

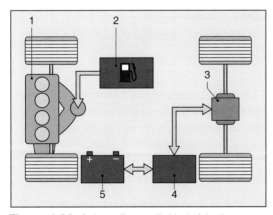

Figure 4.26 Axle-split parallel hybrid: 1) engine, 2) fuel tank, 3) motor, 4) inverter, 5) battery

A conventional transmission is not needed, so this creates space for packaging the overall system – a larger battery, for example. The engine can be optimised to only operate in a set range of rpm. Stopping and starting the engine has no effect on the vehicle drive; therefore, the control systems are less sophisticated. The main disadvantage is that the energy has to be converted twice (mechanical to electrical, and electrical back to mechanical), and if the energy is also stored in the battery, three conversions are needed. The result is decreased efficiency, but this is made up for by operating the engine at its optimum point. There is a 'packaging advantage' in this layout because there is no mechanical connection between the engine and the wheels.

Definition
A series hybrid configuration is a layout where the engine drives a generator (alternator) that charges the battery that powers a motor that drives the wheels.

This configuration tended to be used in trains and large buses rather than cars. However, this layout is now used for range-extended electric vehicle (REVs). In this case, the car is effectively pure electric, but a small engine is used to charge the battery and 'extend the range' or at least reduce range anxiety.

Series-parallel hybrid systems are an extension of the series hybrid because of an additional clutch that can mechanically connect the generator and motor. This eliminates the double energy conversion except at certain speed ranges. However, the 'packaging advantage' of the series drive is lost because of the mechanical coupling. Further, two electric units are required compared with the parallel hybrid.

The power-split hybrids combine the advantages of series and parallel layouts but at the expense of increased mechanical

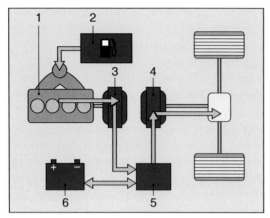

Figure 4.27 Series hybrid: 1) engine, 2) fuel tank, 3) alternator/generator 4) motor, 5) inverter, 6) battery

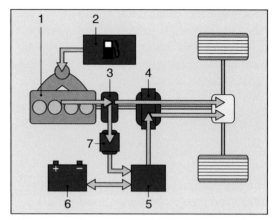

Figure 4.29 Power-split hybrid (single-mode concept): 1) engine, 2) fuel tank, 3) planetary gear set, 4) motor, 5) inverter, 6) battery, 7) generator

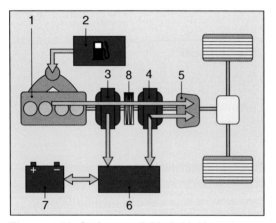

Figure 4.28 Series-parallel hybrid: 1) engine, 2) fuel tank, 3) alternator/generator 4) motor, 5) transmission, 6) inverter, 7) battery, 8) clutch

complexity. A proportion of the engine power is converted to electric power by the alternator and the remainder, together with the motor, drives the wheels. A power-split hybrid is a strong hybrid because it meets all the required functions.

The single-mode concept shown in Figure 4.29 uses one planetary gear set (a dual-mode system uses two and can be more efficient but even more complex mechanically). The gear set is connected to the engine, alternator

and the motor. Because of the epicyclic gearing, the engine speed can be adjusted independently of the vehicle speed (think of a rear-wheel-drive differential action where the two halfshafts and propshaft all run at different speeds when the car is cornering). The system is effectively an electric constantly variable transmission (eCVT). A combination of mechanical and electrical power can be transmitted to the wheels. The electrical path can be used at low power requirements and the mechanical path for higher power needs.

Definition

eCVT: electric constantly variable transmission.

The system, therefore, achieves good savings at low and medium speeds but none at high speeds, where the engine only drives mostly via the mechanical path.

4.2.5 48 V hybrid system

Bosch has developed a hybrid powertrain that makes economic sense even in smaller vehicles. The system costs much less than

Figure 4.30 Power-split hybrid (Source: Toyota)

normal hybrid systems but could still reduce consumption by up to 15%. The electrical powertrain provides the combustion engine with an additional 150 Nm of torque during acceleration. That corresponds to the power of a sporty compact-car engine.

Unlike conventional high-voltage hybrids, the system is based on a lower voltage of 48 V and can, therefore, make do with less expensive components. Instead of a large electric motor, the generator has been enhanced to output four times as much power. The motor generator uses a belt to support the combustion engine with up to 10 kW. The power electronics forms the link between the additional low-voltage battery and the motor generator. A DC/DC converter supplies the car's 12 V 48 V vehicle electrical system. The newly developed lithium-ion battery is also significantly smaller.

The 48 V hybrid systems are a stop-gap technology to improve the emission performance of existing internal combustion engines. However, this is a useful and interesting technology. The 48 V system is not

Figure 4.31 Some 4 million new vehicles worldwide were equipped with a low-voltage hybrid powertrain in 2020 (Source: Bosch Media)

classed a high-voltage. but nonetheless, the voltage could cause a very high current to flow, and strong magnets and high torques are all part of the mix – so, in other words, it should be treated just like any other HEV/EV system.

The Audi A8 – mild hybrid electric vehicle (MHEV) is a good example of 48 V technology. MHEV is standard with all A8

73

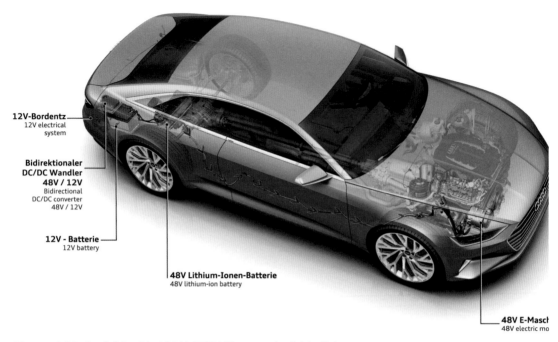

12V-Bordentz
12V electrical
system

**Bidirektionaler
DC/DC Wandler
48V / 12V**
Bidirectional
DC/DC converter
48V / 12V

12V - Batterie
12V battery

48V Lithium-Ionen-Batterie
48V lithium-ion battery

48V E-Masch
48V electric mo

Figure 4.32 Audi A8 with 48 V MHEV (Source: Audi Media)

engines. It reduces fuel consumption by up to 0.7 litres (0.2 gal) per 100 kilometres (62.1 mi) in actual driving. The MHEV technology from Audi is based on a newly developed 48 V main vehicle electrical system. It comprises a compact lithium-ion battery with an electrical capacity of 10 Ah, located in the luggage compartment underneath the loading floor, and a belt alternator starter (BAS or sometimes BSG – originally known as belt starter generator) connected to the crankshaft.

In the speed range of 55 to 160 km/h (34.2–99.4 mph), the A8 can coast with the engine off, if the driver releases the accelerator. The vehicle can then travel with zero emissions for up to 40 seconds. As soon as the driver steps on the pedal again, the BAS prompts a swift, very smooth restart. The new 48 V system allows a high recuperation power of up to 12 kW plus start-stop operation from 22 km/h (13.7 mph).

In combination with the efficiency assist, the adaptive cruise assist slows and accelerates the Audi Q8 predictively by analysing sensor information, navigation data and traffic signs. It automatically adjusts to the current speed limit, reduces the speed before corners, during turning and on roundabouts. The system always considers a driving style in keeping with the selected driving program, from efficient to sporty, and thus, with its predictions, optimally supports the advantages of the mild hybrid technology (MHEV).

If the adaptive cruise assist is deactivated, the efficiency assist displays predictive tips in the instrument cluster and also provides haptic feedback through the active accelerator pedal. By so doing the system helps the driver to drive efficiently. This results from the use of the MHEV recuperation and the intelligent selection of coasting or thrust, depending on the events ahead.

4.2.6 Porsche 800 V performance battery system

The battery is located in the underbody of the Porsche Taycan, ensuring a low centre of gravity and thus sporty driving characteristics. Its housing is a load-bearing component of the body structure, accommodating cooling and electronic components and protecting them from environmental influences.

The waterproof housing is a sandwich construction consisting of a cover at the top and a bulkhead plate at the bottom. The truss-design battery frame with multiple subdivisions is mounted in-between. The cooling elements are glued on underneath the bulkhead plate. The battery housing is secured by means of a steel protective plate. For the battery frame, the developers opted for a lightweight aluminium design. On the one hand, this provides a lot of installation space for the cell modules – and consequently a high battery capacity. On the other hand, this has made it possible for the

vehicle weight to be kept low. Modern joining techniques are used such as MIG welding (metal welding with inert gases) on the battery frame, laser welding on the bulkhead and protection plates and heat-conducting adhesive on the line system under the battery.

The Porsche Taycan is the first production vehicle with a system voltage of 800 volts instead of the usual 400 volts for electric cars. This enables consistent high performance, reduces the charging time and decreases the weight and installation space of the cabling.

The two-tier Performance Battery Plus used in the Porsche Taycan Turbo S and Porsche Taycan Turbo contains 33 cell modules consisting of 12 individual cells each (396 in total). The total capacity is 93.4 kWh. The cells themselves are so-called pouch cells. In this cell type, the stack of electrodes is not enclosed by a rigid housing but by a flexible composite foil. This allows optimal use to be made of the rectangular space available for the battery and a reduction in weight.

Figure 4.33 Porsche Taycan Turbo S: Performance Battery Plus with 93.4 kWh, 800 V system voltage: Less weight, faster charging (Source: Porsche Media)

The modules each have an internal control unit for monitoring voltage and temperature and are connected to each other via busbars. The foot garages – recesses in the battery in the rear footwell – provide the best possible seating comfort in the rear and allow the low vehicle height typical of sports cars.

The battery is integrated into the vehicle's cooling circuit (Figure 4.34) via a line system and a coolant pump. It can be cooled or heated so that it always operates in an ideal temperature window. The cooling elements have been placed outside the actual battery box and are glued to its underside so as to allow heat transfer. The fundamental development aim was to dissipate as little heat as possible into the environment and, thus, be as energy efficient as possible in winter.

The battery (Figure 4.35) can also store the waste heat from the liquid-cooled high-voltage components. As a result, it serves as a thermal storage device or buffer, which permits intelligent functions such as conditioning to ensure driving performance: The target temperature of the battery is determined on the basis of the battery charge and the selected driving program. This ensures sporty driving performance and allows the *Launch Control* feature to be used.

Depending on the outside temperature, the battery is preconditioned to a certain temperature level when the vehicle is connected to the mains for charging. The interior can be preconditioned independently of the mains.

The vehicle also predicts the electrical power consumption of the air-conditioning system and the conditioning of the components based on the outside temperature, humidity and sunshine as well as the currently selected driving program and the respective setting of the automatic climate control system. The current range is calculated using these figures. In a parallel process, Porsche Intelligent Range Manager (PIRM) provides a background forecast for the other driving programs. When route guidance is activated, if the range calculation shows that the destination can be reached with a low battery charge, the system switches to a more energy-efficient driving

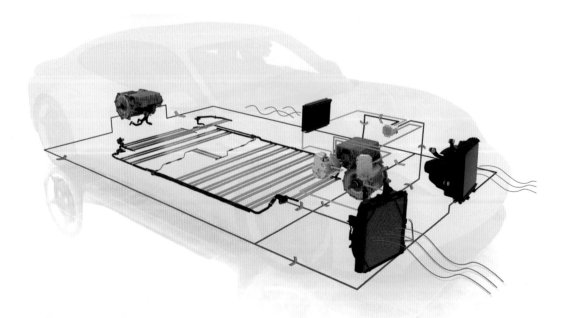

Figure 4.34 Heat pump permits intelligent functions (Source: Porsche Media)

Figure 4.35 Active cooling for battery and electric motors
(Source: Porsche Media)

programme and a different climate-control mode.

4.2.7 Hybrid control systems

The efficiency that can be achieved with the relevant hybrid drive is dependent on the hybrid configuration and the higher-level hybrid control. Figure 4.36 uses the example of a vehicle with a parallel hybrid drive. Shown are the networking of the individual components and control systems in the drivetrain. The higher-level hybrid control coordinates the entire system, the subsystems of which have their own control functions. These are:

▶ battery management
▶ engine management
▶ electric drive management
▶ transmission management
▶ braking system management

In addition to control of the subsystems, the hybrid control also includes an operating strategy that optimises the way in which the drivetrain is operated. The operating strategy directly affects the consumption and emissions of the hybrid vehicle. This is during stop/start operation of the engine, regenerative braking and hybrid and electric driving.

4.2.8 Hybrid case study

Lamborghini's first high-performance electrified vehicle (HPEV) hybrid super sports car was scheduled to make its debut in 2023.

The car (known as LB744) uses a new layout combined with a new powertrain to deliver over 1,000 ps (986.32 hp) in total. This is done by combining a new 12-cylinder internal combustion engine and three electric motors. A double-clutch gearbox is also employed.

Lamborghini has chosen a naturally aspirated 6.5 L V12 mid-mounted engine, which is assisted by three electric motors. One of these motors is integrated into the double-clutch 8-speed gearbox, which is mounted transversely behind the combustion engine. The transmission tunnel houses a 3.8 kWh lithium-ion battery. The car can be used as

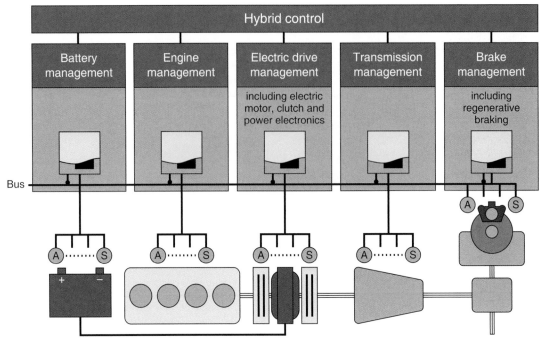

Figure 4.36 Parallel hybrid control system. A, actuator; S, sensor

Figure 4.37 Power components and part of the V12 engine (Source: Lamborghini Media)

a pure electric vehicle, reducing overall CO_2 emissions by 30% compared to a recent Lamborghini ICE-only super sports car.

The V12 engine weighs just 218 kg, has a compression ratio of 12.6:1 and a maximum power output of 825 ps at 9,250 rpm. Maximum torque is 725 Nm at 6,750 rpm. The air intake ducts have been redesigned and can handle an increased amount of airflow. Combustion within the engine

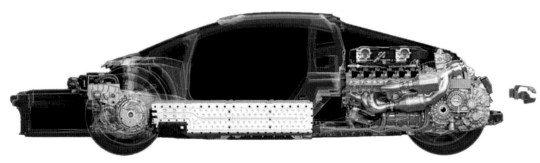

Figure 4.38 Component location (Source: Lamborghini Media)

Figure 4.39 V12 engine and front drive motors (Source: Lamborghini Media)

has also been optimised by regulation of ionisation.

The car has a four-wheel drive system, with the ICE providing power to the rear wheels, while a pair of electric motors supply power to each of the front wheels. As previously mentioned, the third electric motor situated above the gearbox can also supply power to the rear wheels. The drive configuration depends on the selected driving mode. Each front-drive electric motor can deliver 350 Nm and features a torque-vectoring function to optimise driving dynamics. They are used for recuperating energy under braking.

A wet double clutch was chosen as the most efficient and performance-oriented solution. It is capable of handling the high torque figures from the internal combustion engine. Inside the newly developed gearbox are two distinct shafts as opposed to the normal three. One is responsible for the even-numbered gears and the other, the odd gears. Both engage the same rotor, reducing overall weight and saving space.

The rear electric motor, with a maximum power of 110 kW and a peak torque of 150 Nm, works as both the starter motor and generator. It also supplies energy to the front electric motors via the battery in the transmission tunnel. In full electric mode, the motor provides power to the rear wheels, while the e-motors at the front drive the front wheels. This allows pure EV and four-wheel drive. Depending on the selected driving mode, an uncoupling mechanism is used with a synchroniser to enable a connection to the double-clutch gearbox.

4.2.9 Efficiency

The big E word is Efficiency! This is where engineers spend most of their time making small but important steps that improve operation of a vehicle. Efficiency is the ratio of the useful work performed by a machine or process compared to the total energy consumed. It can be expressed as a percentage, but engineers also use the Greek letter Eta (η) as the symbol for efficiency.

> **Definition**
> Efficiency: The ratio of the useful work performed by a machine or process compared to the total energy consumed.

Figure 4.40 shows the main types of hybrid and notes some typical transmission efficiency figures:

- ▶ P0: e-machine is connected to the crankshaft by belt
- ▶ P1: e-machine is assigned to the transmission input shaft between ICE and coupling
- ▶ P2: e-machine is assigned to the transmission input shaft. (P2.5: e-machine is integrated in the hybrid transmission)
- ▶ P3: e-machine is assigned to the transmission output shaft
- ▶ P4: e-machine is integrated in the axle drive

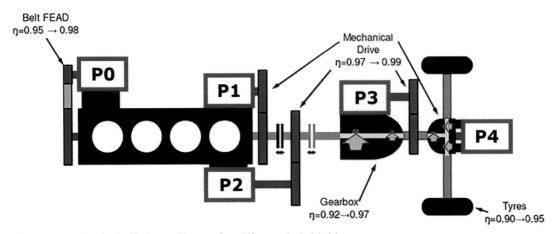

Figure 4.40 Typical efficiency figures for different hybrid drives

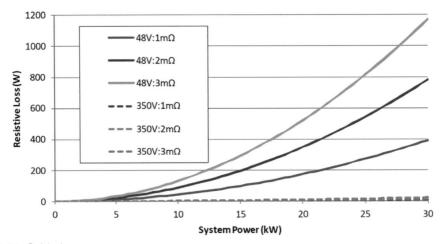

Figure 4.41 Cable losses

In addition to the mechanical efficiency of various drives, it is important to look at the figures for the battery, motor/generator and power transmission (cabling). The way the battery is constructed has an impact on efficiency too. An interesting comparison is between a 350 V system and a 48 V system.

The battery will have a typical efficiency of 0.83 to 0.91 at 350 V, and 0.80 to 0.88 at 48 V. However, at lower voltage and, therefore, higher current, the cabling power loss increases. Cables have a typical impedance of 1.5 to 3 mΩ. Figure 4.41 shows typical power loss in cables and how it is greater at lower voltage.

However, a 48 V P0 mounted machine can still offer similar benefit to a P2–P4 350 V system (Figure 4.42). The aspects compared are

▶ total battery capacity for a mild hybrid application typically 500 Wh to 1 kWh
▶ useable energy requirements typically 150–200 Wh
▶ power/energy and, hence, cell chemistry requirements are the same whether battery is 48 V or 350 V
▶ 350 V or 48 V hybrid battery losses are similar but interconnection losses can be higher with 48 V due to more current flowing
▶ 48 V battery potentially more reliable as one open circuit cell only has minimal effect

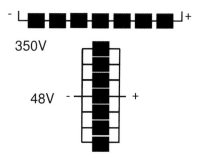

Figure 4.42 Series and parallel cell blocks and 48 V or 350 V connections

Figure 4.44 Orange cables and warning stickers on a Golf GTE

A 48 V P0 system is less complex and lighter than the higher voltage system but has an upper power limit.

4.3 Cables and components

4.3.1 High-voltage cables

Any cable used on a vehicle should be insulated to prevent contact and short circuits. Most cables are made from many strands of copper wire as this offers low resistance and retains flexibility. The insulation is normally a form of PVC.

High-voltage cables require greater insulation to prevent voltage leakage but also because the risk of harm if touched is very high. Stickers with various symbols are used as a warning together with the bright orange colour.

> **Safety First**
> Stickers with various symbols are used as a warning together with the bright orange colour.

To deliver high power, they have to carry high current – even at high-voltage! Remember, power equals voltage multiplied by current ($P = IV$). Current, therefore, equals power divided by voltage ($I = P/V$). We will assume a voltage of 250 V to make the calculations easy! If a cable has to deliver, say, 20 kW (20,000 W) then 20,000/250 = 80 A. Under hard acceleration, this figure is even higher: 80 kW for example would require a current of 320 A. For this reason, the cables are quite thick as well as well-insulated.

Figure 4.43 VW Golf-e showing some of the orange cables

Figure 4.45 Toyota Prius under bonnet view

> **Definition**
> Power equals voltage multiplied by current (P = IV).

4.3.2 Components

It is important to be able to identify EV components. In many cases the manufacturer's information will be needed to assist with this task. Slightly different names are used by some manufacturers, but, in general, the main components are

- battery
- motor
- relays (switching components)
- control units (power electronics)
- charger (on-board)
- charging points

- isolators (safety device)
- inverter (DC-to-DC converter)
- battery management controller
- ignition/key-on control switch
- driver's display panel/interface

Some of these components are also covered in other parts of this book. Possible additions to this list are other vehicle systems such as braking and steering or even air conditioning as they have to work in a different way on a pure EV. The key components will now be described further.

Battery: The most common battery technology now is lithium-ion. The complete battery pack consists of a number of cell modules (the actual battery consisting of 200–300 cells), a cooling system, insulation, junction box, battery management and a suitable case or shell. These features combine so that the pack

Figure 4.46 High-voltage components shown in red, braking components in blue, low-voltage in yellow and sensor/date shown in green (Source: Bosch Media)

is able to withstand impacts and a wide range of temperatures.

> **Key Fact**
>
> The most common battery technology now is lithium-ion.

The battery is usually installed in the underbody of the car. On a pure EV, it can weigh in excess of 300 kg and, for a PHEV, in the region of 120 kg. Voltages vary and can be up to 800 V; however, typically this is around 400 V. The capacity of the battery is described in terms of kilowatt-hours and will be in the region of 20–25 kWh.

> **Key Fact**
>
> The kilowatt-hour (kWh) is a unit of energy equal to 1000 Wh or 3.6 MJ. It is used to describe the energy in batteries and as a billing unit for energy delivered by electricity suppliers. If you switch on a 1 kW electric fire and keep it on for 1 hour, you will consume 1 kWh of energy.

Battery management controller: This device monitors and controls the battery and determines, among other things, the state of charge of the cells. It regulates the temperature and protects the cells against overcharging and deep discharge. Electronically activated switches are included that disconnect the battery system when idle and in critical situations such as an accident or fire. The device is usually part of the battery pack – but not always, so check the manufacturer's data.

Motor: This is the component that converts electrical energy into kinetic energy or movement – in other words, it is what moves the vehicle. Most types used on EVs, HEVs and PHEVs are a type of AC synchronous motor supplied with pulses of DC. They are rated in the region of 85 kW on pure EVs.

Inverter: The inverter is an electronic device or circuit that changes direct current (DC) from the battery to alternating current (AC) to drive the motor. It also does this in reverse for regenerative charging. It is often described as the power electronics or similar. Sometimes the same or a separate inverter is used to supply the 12 V system.

Control unit: Also called power control unit or motor control unit, this is the electronic device that controls the power electronics (inverter). It responds to signals from the driver (e.g., brake, acceleration) and causes the power electronics to be switched accordingly. The control makes the motor drive the car or become a generator and charge the battery. It can also be responsible for air conditioning, power-assisted steering and brakes.

Charging unit: This device is used on pure EVs and PHEVs and is usually located near where the external power source is connected. It converts and controls the 'mains' voltage (typically 230/240 V AC in Europe and 120 V AC in the USA) to a suitable level for charging the battery cells (typically 300 V DC).

Driver interface: To keep the driver informed, a number of methods are used. Most common now is a touch-screen interface where information can be delivered as well as allowing the driver to change settings such as the charge rate.

4.3.3 Inverter developments

Inverters are a critical part of the electric vehicle drivetrain. The inverter controls when and where current flows in the motor to efficiently control motor torque and speed. All the electrical power and energy delivered to the motor must first flow through the inverter, so inverter efficiency is extremely important to overall vehicle efficiency to maximise vehicle range. Ricardo is an important company working on developments in this area.

From a technical standpoint, silicon-carbide (SiC) power electronics development is

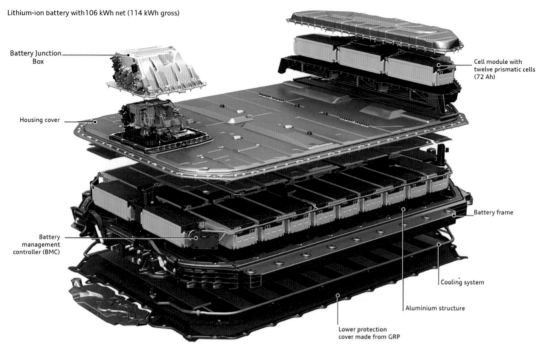

Lithium-ion battery with 106 kWh net (114 kWh gross)

Battery Junction Box

Cell module with twelve prismatic cells (72 Ah)

Housing cover

Battery frame

Battery management controller (BMC)

Cooling system

Aluminium structure

Lower protection cover made from GRP

Figure 4.47 Battery pack (Source: Audi Media)

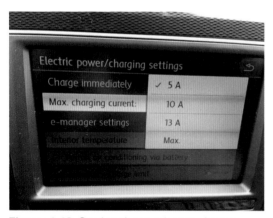

Figure 4.48 Setting the maximum charge current (PHEV)

focused around reducing the size and weight of systems. This will include advancements in lowering power loss, high frequency operation, higher junction temperature and higher voltage operation. Operating an electric vehicle at higher efficiency translates to greater range.

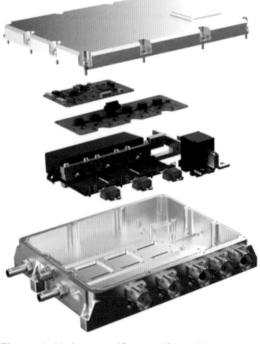

Figure 4.49 Inverter (Source: Ricardo)

Innovation around inverter technology has been advancing rapidly. The technology has evolved from insulated gate bipolar transistors (IGBTs) to silicon carbide as a response to size and weight considerations. SiC inverters provide reduced operational costs, are smaller and lighter, and are more efficient because they can handle higher voltage. This, in turn, allows vehicles to be lighter and increase charging rates. Advancements in the electrification architectures are driving systems upwards to improve vehicle performance compared to 400 V through greater range, higher charging rates and lower vehicle weight. Therefore, 800 V and higher architectures results in greater vehicle efficiency.

Over the next decade, innovations in SiC will continue to focus on overall system weight and size reductions to achieve even greater power density as well high(er) efficiency for overall improved system operational costs. At the power module level, the technology is now reliable and efficient, particularly at high-voltages >600 V where it is unmatched, though gallium nitride (GaN) technology is catching up and providing a compelling alternative for applications <600 V and <100 kW, where it may continue to be a more manufacturable and cost-favourable alternative.

In summary, SiC inverters are smaller and lighter, enabling engineers and designers to take advantage of aerodynamic and packaging improvements and reducing the amount of cooling a vehicle requires.

4.3.4 ECE-R100

ECE-R100 is a standard developed by the UN to harmonise EV systems worldwide. It is applicable for EVs, vehicle category M and N, capable of a top speed above 25 km/h. In this section, we have highlighted some key aspects of the regulation. It is generally about safety of the high-voltage parts in an EV.

Review the document *Uniform provisions concerning the approval of vehicles with regard to specific requirements for the*

electric power train: https://unece.org/ transport/documents/2022/03/standards/ regulation-no-100-rev3

> **Definition**
> ECE-R100 is a standard developed by the UN to harmonise EV systems.

> **Key Fact**
> *Category M: Motor vehicles with at least four wheels designed and constructed for the carriage of passengers.*
> *Category N: Motor vehicles with at least four wheels designed and constructed for the carriage of goods.*

Protection against electrical shock is a key aspect of the standard:

▶ It should not be possible that live, high-voltage parts in passenger and luggage compartments can be touched with a standardised test-pin or test-finger.
▶ All covers and protection of live high-voltage parts should be marked with the official symbol (Figure 4.50) and access

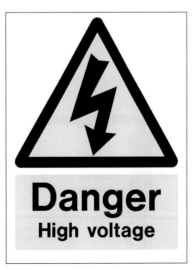

Figure 4.50 This warning symbol may be used with or without the text

to live high-voltage parts should only be possible by using a tool and on purpose.

▶ Traction battery and powertrain shall be protected by properly rated fuses or circuit breakers.

▶ The high-voltage powertrain must be isolated from the rest of the EV.

Charging:

▶ The EV should not be able to move during charging.

▶ All parts that are used while charging should be protected from direct contact, under any circumstance.

▶ Plugging in the charging cable must shut the system off and make it impossible to drive.

General safety and driving points:

▶ Starting the EV should be enabled by a key or suitable keyless switch.

▶ Removing the key prevents the car being able to drive.

▶ It should be clearly visible if the EV is ready to drive (just by pushing the throttle).

▶ If the battery is discharged, the driver should get an early warning signal to leave the road safely.

▶ When leaving the EV, the driver should be warned by a visible or audible signal if the EV is still in driving mode.

▶ Changing the direction of the EV into reverse should only be possible by the combinations of two actuations or an electric switch that only operates when the speed is less than 5 km/h.

▶ If there is an event, like overheating, the driver should be warned by an active signal.

4.4 Other systems

4.4.1 Heating and air conditioning

Most EVs now allow the operation of heating or cooling when the vehicle is plugged in and charging. Some also allow this function directly from the battery. The most important aspect is that this allows the vehicle cabin to be 'pre-conditioned' (heated or cooled) on mains power, therefore saving battery capacity and increasing range. The two most common systems allow

▶ cooling with an electrically driven air conditioner compressor

▶ heating with a high-voltage positive temperature coefficient (PTC) heater

These cooling and heating functions using the high-voltage components are usually activated with a timer or a remote app.

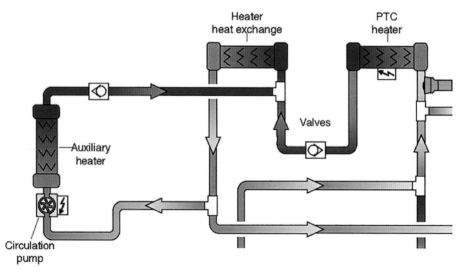

Figure 4.51 Heating circuit (Source: Volkswagen)

Key Fact

Most EVs now allow the operation of heating or cooling when the vehicle is plugged in and charging.

Hybrid car systems combine the heating circuit by running it in parallel to the coolant circuit. It consists of a heat exchanger, a heater unit and a feed pump.

Cooling systems operate in much the same way as on a conventional vehicle, except that the compressor is electrically driven. This can be by high-voltage or lower-voltage systems such as 42 V (but not normally from 12 V).

When necessary, the battery control unit can request cooling of the battery when it is being charged, so, for this reason, the battery cooling circuit and, in some cases, the motor cooling circuit are combined with the engine cooling system on a hybrid. The electric pump makes the coolant flow.

4.4.2 Thermal challenges

Battery electric vehicle (BEV) supporters are keen to point to the 95% efficiency of the electric motor. This is compared with the most advanced petrol/gasoline engines, which have 40% thermal efficiency. These numbers can, however, be misleading. The figure for petrol engines is under high-load conditions. The number is lower, more like 30%, under other operating conditions. BEVs also suffer from electrothermal losses.

However, engineers are looking at how to use all waste heat to warm passengers in cold conditions.

4.4.3 Liquid cooling

The operation of the DC–DC converter, power electronics and battery charging/discharging cycles all produce heat (Table 4.3). The electrical energy losses may total up to 40%.

There is, therefore, a need for liquid cooling (also see Section 5.3.2).

Table 4.3 Thermal losses (increased charge and discharge power significantly increases waste heat)

Component	Thermal loss %	Temperature limit °C
Motor/generator	3–5	85
Power electronics	3–5	65
High-voltage battery	5–8	50
DC–DC converter	3–5	70
Charger	3–5	65

Electric propulsion appears to have a significant efficiency advantage compared to ICE because of losses in the form of waste heat. However, the comparison is closer when the energy to produce the electricity is considered.

The size and weight of the lithium-ion battery pack are significant contributors to the cost of a BEV. There are also the resulting issues of range, recharging time and the deteriorating effects on lithium-ion batteries in high-voltage, high-rate charging. In due course, solid-state batteries could open the path to alternative chemistries with lower flammability than current lithium-ion batteries. For the same energy density, the solid-state type would be more compact and lighter. In the meantime, we have to work with what we have.

Key Fact

The size and weight of the lithium-ion battery pack are significant contributors to the cost of a BEV.

The current focus is on liquid cooling as an ideal way to deal with the challenges of high-voltage, high-rate charging and greater battery density. Air cooling is adequate for the smaller battery packs in hybrids. Liquid cooling permits controlled use of waste heat. Using this heat energy is better than adding expensive battery capacity or even a heat pump, which is a costly addition. Liquid cooling is also ideal for power electronics waste heat management and recovery.

Battery refrigeration systems are under development for future BEVs so they can withstand rapid charging of 100 kW or more. Continental is working on an interesting system known as an electrothermal recuperator (ETR). This is a liquid cooling system for a BEV. The ETR is designed to collect waste heat from regenerative braking. The size of a battery pack and its state of charge can prevent full use of available regenerative braking energy in the form of recharging the battery. Low ambient temperatures can also be an issue.

In this system, when the battery pack is unable to accept charge, the computer controlled ETR circuit causes the coolant to be heated and used for cabin warmth. The ETR system weighs about 3 kg, has a CAN interface for computer control and a built-in PTC resistance heater for additional cabin warming if needed. Based on simulations of a car with a 64 kWh battery pack, potential energy savings were between 5% and 12%.

4.4.4 Heat pumps

A heat pump (Figure 4.52) transfers heat energy from a source of heat to a thermal reservoir. It moves thermal energy in the opposite direction of spontaneous heat transfer. It does this by absorbing heat from a cold space and releasing it to a warmer one.

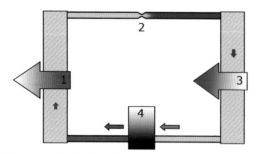

Figure 4.52 A simplified diagram of a heat pump's vapor-compression refrigeration cycle: 1) condenser, 2) expansion valve, 3) evaporator, 4) compressor

> **Key Fact**
> A heat pump transfers heat energy from a source of heat to a thermal reservoir.

A heat pump uses external power to accomplish the work of transferring energy from the heat source to the heat sink. The most common design of a heat pump is almost exactly like a car AC system, working in reverse. It includes a condenser, an expansion valve, an evaporator and a compressor. The heat transfer medium is a refrigerant. The key process is that evaporation of the refrigerant draws heat from its surroundings. Lick the back of your hand and blow on it! It feels cool because the moisture draws heat from your body as it evaporates.

Heat pumps usually can be used either in heating mode or cooling mode, as required by the user. When a heat pump is used for heating, it employs the same basic refrigeration cycle used by an air conditioner or a fridge but in reverse. It releases heat into the conditioned space rather than the surrounding environment. In-car heat pumps draw heat from the external air even when it is cooler than the cabin.

Heat pumps are significantly more energy-efficient than simple electrical resistance heaters. However, the cost of installing a heat pump is much higher than a PTC heater.

4.4.5 Brakes

Brakes are normally operated hydraulically but with some sort of servo assistance. This can be from a hydraulic pump, but on most ICE-driven vehicles, the vacuum (low pressure) from the inlet manifold is used to operate a servo. On a pure EV or a hybrid running only on electricity, another method must be employed.

> **Key Fact**
> Brakes are normally operated hydraulically but with some sort of servo assistance.

In most cases, an electrically assisted master cylinder is used, which also senses the braking pressure applied by the driver. The reason for this is that as much braking effect as possible is achieved through regeneration because this is the most efficient method. The signals from the master cylinder sensor are sent to an electronic control system, and this, in turn, switches the motor to regenerative mode, charging the batteries and causing retardation, or regenerative braking. If additional braking is needed, determined by driver foot pressure, the traditional hydraulic brakes are operated with electrical assistance, if needed.

the braking command, and an integrated pedal travel simulator ensures the familiar pedal feel. The braking pressure modulation system implements the braking command using the electric motor and wheel brakes. The aim is to achieve maximum recuperation while maintaining complete stability. Depending on the vehicle and system status, deceleration of up to 0.3 g can be generated using only the electric motor. If this is not sufficient, the modulation system uses the pump and high-pressure accumulator.

Figure 4.53 iBooster – electronically controlled brake master cylinder (Source: Bosch Media)

Figure 4.54 Vacuum-independent braking system specially designed for plug-in hybrids and electric vehicles. It comprises a brake operation unit (left) and an actuation control module (right), which supplement the ESP® hydraulic modulator (Source: Bosch Media)

Some braking systems have a feedback loop to the master cylinder to give the driver the appropriate feel from the brake pedal that is related to the amount of retardation overall (friction brakes and regenerative brakes).

Not yet in use but coming soon . . . a fully hydraulic actuation system (HAS) has been developed by Bosch for use on electric and hybrid vehicles. The system is suitable for all brake-circuit splits and drive concepts. It comprises a brake operation unit and a hydraulic actuation control module, which supplement the ESP® hydraulic modulator. The brake pedal and wheel brakes are mechanically decoupled. The brake actuation unit processes

4.4.6 Power-assisted steering

When running on electric-only or if there is no engine available to run a power-steering pump, an alternative must be used. However, most modern ICE vehicles use one of two main ways to use electric power-assisted steering (ePAS); the second of these is now the most common by far:

1. An electric motor drives a hydraulic pump, which acts on a hydraulic ram/rack/servo cylinder.
2. A drive motor, which directly assists with the steering.

With the direct-acting type, an electric motor works directly on the steering via an epicyclic gear train. This completely replaces the hydraulic pump and servo cylinder.

> **Key Fact**
>
> With a direct ePAS, an electric motor works directly on the steering via an epicyclic gear train.

Figure 4.55 Electric PAS (Source: Bosch Media)

On many systems, an optical torque sensor is used to measure driver effort on the steering wheel (all systems use a sensor of some sort). The sensor works by measuring light from an LED, which is shining through holes. These are aligned in discs at either end of a torsion bar, fitted into the steering column. An optical sensor element identifies the twist of two discs on the steering axis with respect to each other, each disc being provided with appropriate codes. From this sensor information, the electronic control system calculates the torque as well as the absolute steering angle.

> **Key Fact**
>
> Electrical PAS occupies little under-bonnet space and typically a 400 W motor averages about 2 A even under urban driving conditions.

4.4.7 DC-to-DC converter

A DC-to-DC converter is a device that converts a source of direct current (DC) from one voltage level to another. In most systems, one DC voltage is converted to AC using an inverter, the voltage of this AC is changed using a transformer, and it is then rectified back into DC.

> **Key Fact**
>
> A DC-to-DC converter is a device that converts a source of direct current (DC) from one voltage level to another.

Electric vehicles use a high-voltage battery (generally 200 to 450 V) for traction and a low-voltage (12 V) battery for supplying all the electric components in the vehicle. On ICE vehicles. the low-voltage battery is charged from an alternator, but in an EV, it is charged from the high-voltage battery. Some hybrid vehicles also allow the low-voltage battery to help recharge the high-voltage pack if the vehicle does not use a starter motor.

STMicroelectronics is a world leader in providing semiconductor solutions. They have a wide offer of discrete semiconductors including IGBTs and both silicon and silicon-carbide (SiC) MOSFETs and diodes. Figure 4.56 shows a bi-directional DC–DC converter.

> **Definition**
>
> SiC: Silicon-carbide.

4.4.8 Thermal management case study

Automotive tech specialist Continental has introduced several thermal management (TM) technologies for increasing EV efficiency during summer and winter. This includes sensors, brushless DC (BLDC)-driven pumps, hydraulics

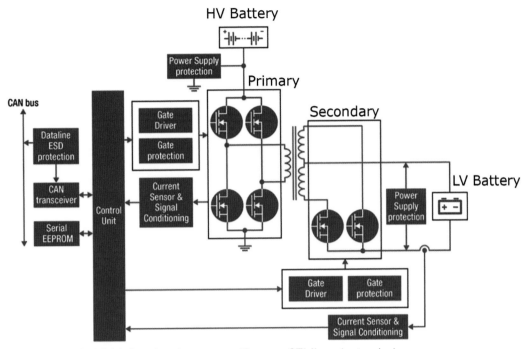

Figure 4.56 DC–DC bi-directional converter (Source: STMicroelectronics)

and valves. Continental claims that its systems can increase EV range by up to 25% in −−10 °C conditions.

Normally, an EV's range can drop by up to 40% at −10 °C as compared to 20–25 °C. Several small changes at these temperatures can mitigate much of this loss. Continental's coolant flow control valves (CFCV) are designed to seamlessly switch between heating and coolant circuits to provide pre-conditioning to battery and drivetrain components when needed. The system can cycle between two and four coolant circuits thanks to its actuator, which uses a compact brushless motor controlled by an intelligent positioning system.

During winter and midsummer, there is a high demand for heating or cooling. For example, at −10 °C, up to 30% of the stored electric energy is needed for heating purposes. A comprehensive thermal management serves to facilitate as much heating and cooling as possible without draining the battery.

Figure 4.57 Thermal management control actuator (Source: Continental AG)

4.4.9 Radiant heating

Vehicle manufacturers have been steadily enhancing the electric range of their vehicles, yet the claimed ranges are typically based on ideal thermal conditions. The true challenge lies in delivering the same performance under

91

extreme thermal conditions that can affect the efficiency of lithium-ion battery packs.

During hot summer weather, using air conditioning in electric vehicles results in a modest reduction in range. However, in cold winter weather, when the cabin heater is in use, there can be a significant drop of 30–40% in range. Until now, most efforts to improve EV heating performance have focused on electric seat heaters, some electric steering wheel heaters and heat pumps integrated into the HVAC refrigeration system. However, heat pump performance is limited by the fact that refrigerant flow slows down in low winter temperatures. As a result, heat pumps must switch to less-efficient resistance or PTC heaters when ambient temperatures drop below freezing.

Toyota's Prius Prime PHEV introduced an innovative solution by incorporating a liquid-gas separator and refrigerant gas injection into its A/C refrigeration system, a technology derived from static-mount commercial heat pumps. This advancement enables the heat pump to operate efficiently even at temperatures as low as −10 °C (14 °F).

In a 2022 presentation at the Toyota showcased its testing of an electric radiant heater installed in a Prius Prime. This system offers a more efficient and rapid method of providing supplementary cabin heat compared to other methods. This development indicates that the Prius Prime, which already features

Figure 4.58 Vehicle interior (Source: Toyota Media)

seat and steering wheel heaters, is being utilised as a testbed for addressing the challenge of maintaining cabin heat while preserving range.

According to Toyota's testing, the use of the radiant heater led to a 5.3% reduction in fuel consumption while significantly improving driver comfort within two minutes of entering the cabin. The radiant heater operates by emitting invisible infrared waves that quickly warm objects in their path. When directed towards the driver in the Prius Prime's cabin, the system warms the front part of the driver's body, complemented by a seat heater covering the back of the body from shoulders to knees. Additionally, a steering wheel heater warms the hands, while the HVAC's floor outlet gradually provides warmth to the driver's knees, feet and the rest of the body. The seat and steering wheel heaters employ conductive heating, where direct contact with a warm surface transfers heat, while the HVAC utilises convective heating through the movement of liquids to a heater core, such as coolant from the engine or the liquid cooling system of an EV battery pack and vehicle drive system.

Unlike a conventional internal combustion engine–powered vehicle's HVAC system, which gradually warms the entire cabin, the HVAC in the Prius Prime focuses on providing warmth primarily to the driver's feet and lower legs up to the knees. While it still takes some time to reach the desired operating temperature, Toyota relies on other electric heaters to bridge the initial gaps. The combustion engine in the Prius Prime also contributes some heat through its cooling system. However, during EV-only operation, additional heat needs to be generated at a faster rate. To address this, Toyota has demonstrated the effectiveness of the electric radiant heater, which is mounted below the steering wheel in the dashboard. Electric vehicles typically employ conventional electric heaters, such as PTC or resistance types, in similar scenarios. However, Toyota's radiant heater offers instant warmth and improved efficiency as it focuses on a specific zone. This technology will be utilised in Toyota's first battery electric vehicle, the 2023 bZ4X, and there are indications that it may also be installed below the glovebox to enhance passenger-side heating.

The research conducted by Toyota validated the benefits of the radiant-heat design in the Prius Prime. The driver's side of the vehicle featured a Denso-developed thermostatically controlled 150 watt heater. A cold-weather test then was performed by a driver trained in comfort-level sensing.

The test involved setting the HVAC to Auto/22 °C (72 °F) while maintaining the chamber temperature at 6.7 °C (19.9 °F). A comparison was made between the HVAC set to the same temperature but with reduced airflow volume and the activation of the radiant heater, which focused on heating the anterior thighs and shins down to the ankles without relying solely on convective heat from the HVAC. The strategy aimed to warm the feet and legs well below the knees for faster engine warm-up, utilising radiant heat primarily for the upper body.

The additional electricity consumed by the radiant heater was determined to be insignificant. Within two minutes of entering the vehicle and activating all systems, the driver in the test vehicle experienced significant comfort, except for the arms, head and feet. In contrast, most of the driver's body without radiant heat remained cold, except for the back and head. Within five minutes, the test driver with the radiant heater engaged was almost fully comfortable, with only the feet, which are typically harder to warm, slightly below the comfort level. On the other hand, without the radiant heater, it took 10 minutes for the driver to reach a barely comfortable level for almost the entire body, with the feet slightly below the comfort level. However, when the radiant heater was engaged, all parts of the driver's body were in the comfort zone

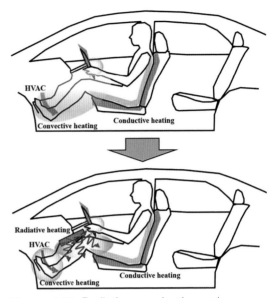

Figure 4.59 Radiative, conductive and convective heating (Source: Toyota Media)

after 10 minutes. By the 20-minute mark, both systems had reached fully comfortable levels.

4.5 High-voltage safety system

4.5.1 Introduction

The high-voltage battery is effectively connected to all the high-voltage components. However, each high-voltage connection can operate a relay. This connects the high-voltage system in the vehicle when the main contactor is closed or disconnects it when it is open (Figure 4.60).

If the contactors are de-energised, they open, and the high-voltage battery is disconnected. The command to open can be triggered by different situations. For example, turning off the vehicle and removing the ignition key opens the contactors. It also activates the other safety systems.

4.5.2 Pilot line

The pilot line is a completely independent safety system that checks if all high-voltage

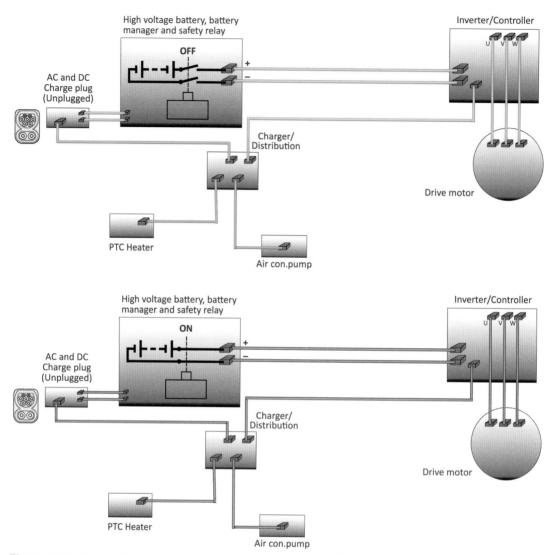

Figure 4.60 High-voltage system components deactivated (OFF) and activated (ON)

components are correctly connected to the high-voltage system. It connects all high-voltage components and operates on low-voltage (Figure 4.61).

The pilot line circuit is interrupted if a high-voltage connection on a high-voltage component is disconnected. This occurs if a cable is disconnected, the maintenance connector is removed or a high-voltage component is replaced.

Key Fact

The pilot line is a completely independent safety system that determines if all high-voltage components are correctly connected to the high-voltage system.

Key Fact

The pilot line circuit is interrupted when a high-voltage contact on a high-voltage component is disconnected.

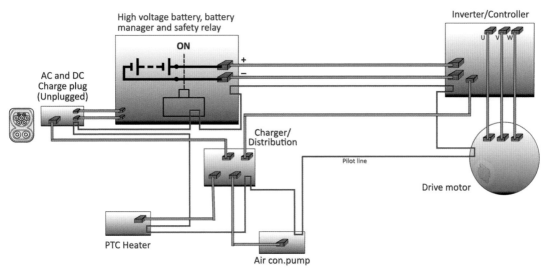

Figure 4.61 The pilot line (in red) runs through every component via their plugs and sockets

The pilot line circuit is a continuous loop. Breaking it at any point causes the protective relays to open and isolate the high-voltage battery.

The high-voltage system normally has a maintenance connector near to the high-voltage battery as an additional safety feature for de-energising the high-voltage system. There may also be a maintenance connector under the bonnet or in other locations (always refer to manufacturer's information). If any connector is unlocked and removed, the pilot line is disconnected and the main contactors open. This disconnects the high-voltage battery, and, in many cases, it also electrically separates the battery into two halves.

The location and appearance of the connector will vary and depends on the vehicle type. When disconnected, the system is de-energised and only the battery modules are live. If electrically halved, the voltage is also halved.

As a reminder, always follow the three basic rules of high-voltage safety:

1. De-energise the high-voltage system.
2. Secure the vehicle against reactivation.
3. Check/determine whether the high-voltage system is de-energised.

De-energisation should only be performed by a qualified person.

4.5.3 Crash safety

The high-voltage safety system is linked to the crash detection system (Figure 4.62), usually via the airbag control module. De-energisation of the high-voltage system protects the vehicle occupants, first responders and technicians working on a vehicle in for repair after an accident.

When the airbag control module detects an accident and deploys the belt tensioner or airbags, the battery regulation control module is instructed, via the CAN data bus, to open the protective relays. There are two scenarios:

▶ single-stage crash deployment: If just the belt tensioners are deployed, the contactor can be closed by turning the ignition on and off again.
▶ second-stage crash deployment: If the belt tensioners and airbags are deployed, the contactors can often only be closed again using the manufacturer's or other suitable equipment.

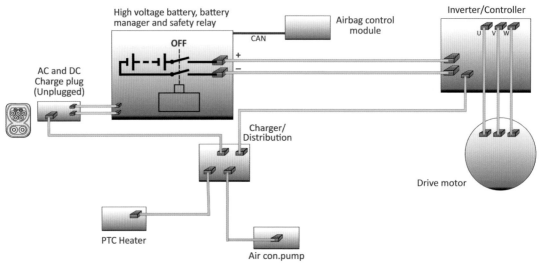

Figure 4.62 Crash detection

4.5.4 **Insulation resistance**

The battery regulation control module transmits a test voltage to check the insulation resistance of the high-voltage system. The voltage is usually about 500 V and has a low current, which is not dangerous. This checks that all the high-voltage components and cables are correctly insulated and shielded. The control module calculates and compares readings with previously measured resistance values of the high-voltage system.

If the insulation of a wire is damaged, for example, by a vermin bite, the insulation resistance changes. The control module detects this as an insulation fault. A range of messages can appear in the vehicle instrument cluster depending on the severity of the fault.

4.6 **Heavy vehicles**

4.6.1 **Overview**

Heavy-vehicle electric drive systems are fundamentally the same as those used on light vehicles, except that they generally need to produce more torque. Some also use more than three phases to drive the motor. The block diagram shown as Figure 4.63 is

a generic layout of a plug-in hybrid electric vehicle (PHEV) or a pure/battery electric vehicle (P/BEV). It could be a light or a heavy vehicle.

The high-voltage battery is the main source of energy and is usually lithium-ion based technology. It is made of many cells in a series and parallel combination to produce a voltage of about 400 V. However, this is tending to increase on newer models, and 800 V or more is likely to become popular. On heavy vehicles, the voltage tends to be higher because more power and, therefore, current needs to be delivered.

> **Key Fact**
>
> On heavy vehicles, the voltage tends to be higher because more power needs to be delivered.

The inverter uses insulated gate bipolar transistors (IGBTs), or similar, to convert DC to three-phase AC, which is used to drive the motor. It is also able to work as a rectifier and take three-phase AC generated in the motor (generator) during braking and change it into DC to recharge the battery. Some systems use a separate electronic control unit (ECU), but many have it integrated into this unit.

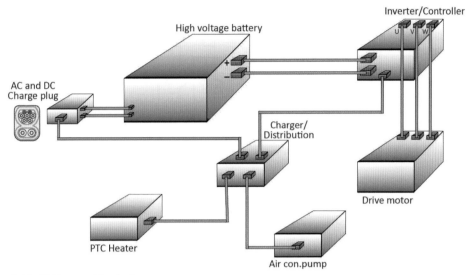

Figure 4.63 EV layout block diagram

The drive motor block in Figure 4.63 is three-phase, and it drives the vehicle or assists with its movement if in hybrid configuration. It also acts as a generator when driven via the wheels and transmission when braking. The charge plug is the type that allows either direct DC charging or normal AC charging. It is plugged into a domestic or more heavy-duty industrial charging unit.

When charging from a normal unit or domestic or industrial mains supplies, the charger unit steps up the voltage (often 230 V AC) to the 400+ V DC needed to charge the battery. It also acts to distribute the higher voltage three-phase AC to items such as the air-conditioning pump and positive temperature coefficient (PTC) heater.

Using a high-voltage air-conditioning pump allows the system to run when either the ICE engine is not operating (on a hybrid) or on a pure electric vehicle.

The heater is effectively an electric fire! It produces heat by using PTC elements in an air flow. PTC elements increase in resistance as temperature increases, so they are self-limiting with respect to current flow.

To operate the motor, different phases are switched on at different times to cause rotation. The signals used to switch the IGBTs are pulse-width modulated for finer control. For some heavy-vehicle applications, multiphase motors are used (Figure 4.64). Arguably a three-phase motor is multiphase, but the term is generally used to mean more than three. The switching of the inverters that drive these motors is more complex but still follows the basic principle of switching a phase in turn. Multiphase drives have some advantages, in addition to their improved power-to-weight ratio, when compared to the standard three-phase versions, for example:

▶ The current stress of the semiconductors is reduced.
▶ Torque ripple is reduced.

Other advantages include a reduction in the noise and reduced stator copper loss, which results in improved efficiency. Some systems are more reliable because even if one or more phases are lost, the motor can still operate with reduced performance. The inverter shown in Figure 4.64 can operate in either three-phase mode or nine-phase mode.

97

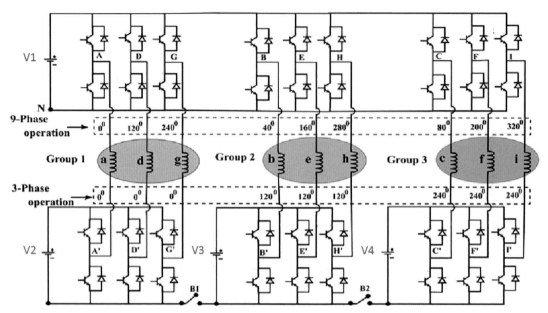

Figure 4.64 Multiphase inverter

The batteries for heavy-vehicle use are the same as for light-vehicle systems but are often made modular so that they can be used at different capacities and voltages. The modular approach also allows easier fitting into different spaces as well as easier repairs when required. Lithium-ion technology is mostly used. Voltages from 400 V to 800 V and more are becoming common.

> **Key Fact**
>
> The batteries for heavy-vehicle use are the same as for light-vehicle systems but are often made modular so that they can be used at different capacities and voltages.

The main components of an EV operate using the same principles, whether used on light or heavy vehicles. On heavy vehicles, however, they will tend to be larger and more powerful simply because of the higher torque and power requirements. Some motors are six- or nine-phase instead of the normal three-phase. This is usually described as multiphase.

4.6.2 Heavy vehicle options

If choosing to operate a heavy EV, there are two important considerations: First, the UK's right-hand-drive market status and its departure from the EU mean that certain vehicle options available in mainland Europe are not yet accessible in the UK. Second, similar to diesel trucks, the range and power consumption of electric trucks are influenced by various external factors. Factors such as the use of electrical accessories (e.g., tail-lifts, cab climate control), ambient temperatures, route characteristics and driver techniques (e.g., freewheeling, regenerative braking) all impact the performance of the vehicle. In practice, these operational variables often have a more significant effect on performance than differences in specifications listed on paper.

Currently, it is challenging to make a strong economic case for directly replacing diesel trucks with electric ones. Operators face high capital costs and operational limitations when transitioning to battery-powered vehicles. Additionally, the relatively expensive electricity prices and the need for suitable charging infrastructure pose further challenges. Only a minority of transport businesses operate from owned depots with sufficient electrical capacity to support an existing electric fleet. Installing expensive charging infrastructure at leased premises is typically not feasible unless landlords provide guarantees or support. It is evident that selecting electric vehicles without adequate infrastructure support is not a prudent decision.

However, there are two factors that can sway the decision in favour of electric trucks. First, when customers are under pressure to reduce their environmental impact, they may be willing to pay a premium for electric options to lower carbon emissions and eliminate toxic fumes. Second, the requirement to enter zero-emissions zones, albeit small in number but growing, can necessitate the use of electric trucks.

When introducing electric trucks, operators must be prepared to redesign their routes to maximise the benefits of regenerative charging. Essentially, trucks should depart from the depot fully charged and tackle the most challenging portions of the route first, allowing them to harvest energy during the return journey, leveraging the principle of uphill out and downhill back.

4.6.3 Bus and coaches

This section will cover some current examples of the components used in coaches and buses. These are considered to be heavy vehicles as well as trucks and construction vehicles.

Canadian company Dana Group produce a family of high-torque electric powertrain systems for electric and hybrid commercial vehicle applications. There are a number of motor models with varying sizes paired with a choice of medium- (<450 V DC) or high-voltage (<750 V DC) inverters.

These high-torque/low-speed permanent magnet motors are designed to interface with standard rear differentials without the need for an intermediate gearbox. By allowing direct drive operation, it reduces powertrain complexity and cost. A direct drive system can produce over 10% efficiency gains throughout the driving cycle, representing an equivalent gain in battery usage and, therefore, range.

Typical applications for these technologies are

▶ city buses
▶ delivery trucks
▶ tow tractors
▶ mining vehicles
▶ marine applications
▶ shuttles

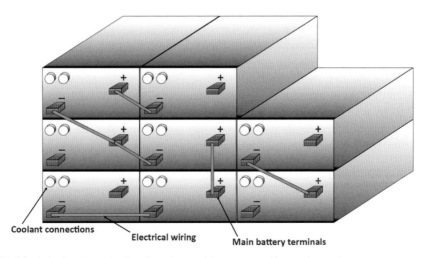

Figure 4.65 Modular battery design (random wiring connections shown)

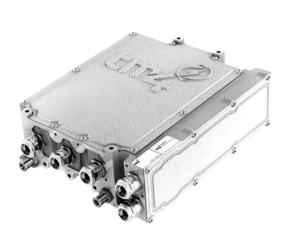

Figure 4.66 Inverter and motor (Source: Dana)

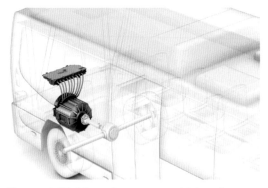

Figure 4.67 Nine-phase motor driving the rear axle of a bus (Source: TM4)

The inverters used for heavy-vehicle applications have to be able to handle significant power levels (voltages and current). Just like on smaller vehicles, these devices convert DC from the battery into three or more phases of AC to drive the motors. They also act to rectify the output of the motor during regenerative braking.

A common approach to increasing power is to use multiple power transistors (IGBTs) in parallel. However, IGBTs are never perfectly matched, and when in parallel, current does not distribute evenly, which can cause a loss in performance of 10% or more compared to the same number of independent IGBTs.

A multiphase topology for inverters and motors uses separate IGBTs to drive independent electromagnetic subsets of the motor. This allows each IGBT to be used to its fullest potential. Because the IGBTs are fully independent, it is possible to use interleaved IGBT switching. This spreads the current ripple demand from the DC bus filtering capacitor among the IGBTs.

Almost all EV motors work on the principle of a permanent magnet rotor and a rotating field stator. Heavy-vehicle systems in this respect are no different. However, the extra power requirements of heavy vehicles mean some innovative technologies have been used. As discussed in the first section, multiphase systems are used, but another interesting method is to mount the rotor outside the stator. This external rotor motor topology has a greater magnetic flux and can produce higher torque. Torque (Nm) is the product of the force (N) and the distance (m) from the centre of rotation. This is represented in Figure 4.69 and shows how the greater distance can result in increased torque.

> **Key Fact**
> An external rotor motor topology has a greater magnetic flux and can produce higher torque.

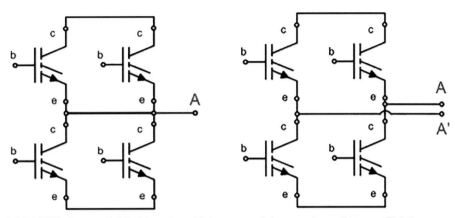

Figure 4.68 IGBTs in parallel (left) and multiphase modular topology (Source: TM4)

Table 4.4 Dana product data (note for comparison, the peak torque for a Tesla Model S is estimated at around 1200 Nm – with two motors)

	SUMO HD HV2700	SUMO HD HV3400	SUMO HD HV3500
Inverter	CO300-HV	CO300-HV	CO300-HV
Peak power (kW)	250	250	350
Continuous power (kW)	195	195	260
Operating speed (RPM)	0–3375	0–2450	3400
Continuous torque (Nm)	2060	2060	1830
Peak torque (Nm)	2700	3400	3500

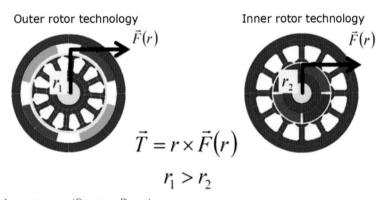

Outer rotor technology   Inner rotor technology

$$\vec{T} = r \times \vec{F}(r)$$

$$r_1 > r_2$$

Figure 4.69 Motor torque (Source: Dana)

The example motor (HV3500) shown in Figure 4.67 is 572 × 591 × 505 mm and weighs 340 kg. The associated inverter is 414 × 126 × 801 mm and weighs in at 36 kg. TM4 offer three main packages (2018) and the specifications are as follows:

The electronic control unit (ECU) runs the hybrid or pure EV system based on operating conditions and driver demand. It is programmed (like all engine or EV control ECUs) to meet the specific needs of the vehicle, operating environment and associated systems.

101

Figure 4.70 The appropriately named 'neuro 200' vehicle management unit is the brain of the system (Source: Dana)

Figure 4.73 Alternative drive axle arrangement (Source: ZF)

Figure 4.71 City bus equipped with ZF technologies (Source: ZF)

Figure 4.72 Drive motors (Source: ZF)

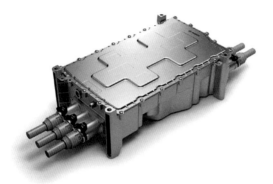

Figure 4.75 eAxle (Source: ZF)

Figure 4.74 HV Inverter (Source: Bosch Media)

In the bus shown in Figure 4.71, two ZF electric motors near the wheels are used (Figure 4.72), each with 120 kW (160 bhp) of maximum power. Like all EV motors, they can also run in generator mode to feed power back into the batteries (regenerative braking).

A key aspect of this system is that it fits into the same space and has the same interface dimensions as portal axles for conventional ICE vehicles. This makes it easy to add to an existing chassis, without needing to redesign the entire vehicle. Shown in Figure 4.73 is

an alternative but similar arrangement of two drive motors.

The ZF truck eAxle (Figure 4.75) offers a compact and integrated solution for electrifying trucks. The eAxle is built as a solid axle and incorporates an electric motor, inverter and transmission within its design. This unit is adaptable to both 400 V and 800 V electrical architectures and provides a power output range of 180 to 350 kW. With a maximum torque output of 15,000 Nm (11,063 lb-ft), it delivers impressive performance.

One notable advantage of this axle is its ability to be seamlessly incorporated into existing and future chassis without extensive engineering modifications.

CHAPTER 5

Batteries

5.1 Overview

5.1.1 Describing a battery

Describing a battery in an EV is more complex that it would first appear. It is important to use the words correctly – although I am sure none of us manage that all the time!

An EV battery is really a battery pack made up of modules, which, in turn, are made up of individual cells. Lithium-ion batteries are the most common by far, but other chemistries are used. Some batteries are cooled directly or indirectly or not at all. Some can be fast-charged and some can't. Their storage capacity is defined as kilowatt-hours (kWh).

We may even mention their energy density (Wh/l), their specific energy (Wh/kg) and specific power (W/kg). Finally, battery cells are ascribed a nominal voltage (V). Remember the importance of temperature too (Figure 5.1).

Store cold, use hot

Charge slowly

Store half full

Figure 5.1 Prolonging the life of a battery (in your phone and car)

DOI: 10.1201/9781003431732-5

5.1.2 **What is a kilowatt-hour?**

The term kilowatt-hour (kWh or kW-h) is often used in connection with electric vehicle batteries but also to household electricity supply. Before I explain in detail what this means, we need to clear up the difference between a kilowatt and a kilowatt-hour. Also, just a reminder, the 'kilo', as I am sure everyone is aware, simply means 1,000. It is used because it makes the numbers easier to work with.

A kilowatt (1,000 watts) is a measurement of *power*. It means how much electricity or horsepower, for example, a device will consume or how much a device can supply. In the context of this article, it relates to electricity, but it doesn't have to. A one horsepower (hp) engine, for example, is about 745 W, or one metric horsepower (Ps) is 735.5 W – I guess, there really is a metric horse out there somewhere! In a vehicle electrical system, an old 60 W headlamp bulb will consume ten times more power than a 6 W sidelight bulb. A 200 kW EV motor will consume considerably more.

A kilowatt-hour (kWh) is a unit of *energy* measurement, specifically it represents the amount of energy used or produced over a period of time. One kilowatt-hour is equal to the energy consumed by a power of one kilowatt over a period of one hour. It is commonly used to describe vehicle traction batteries and as a measure of domestic and industrial electricity consumption. It is the billing unit for electric companies – which is why we have heard so much about it recently!

The following table shows some typical examples of devices with different *power* ratings and their *energy* consumption over time:

A kilowatt-hour is a non-SI unit of energy: one kilowatt of power for one hour. It is equivalent to 3.6 MJ (megajoules) in SI units.

A kilowatt-hour (kWh) is used to measure the energy capacity of electric vehicle batteries because it provides a convenient and standardised way to compare the energy

Table 5.1 Power ratings

Device	Power (kW)	Usage (hours)	Energy (kWh)
Flat screen LED TV	0.1	10	1
AC system	2.5	2	5
Domestic EV charger	3.5	10	35
AC fast charger	22.0	5	110
DC fast charger	60.0	2	120
DC fastest charger*	350.0	20 mins	117

*available at the time of writing

storage capacity of different EV batteries. By expressing the energy storage capacity in kilowatt-hours, manufacturers and consumers can easily compare the range and potential driving distance of different EVs. Note that range of an EV also depends on many other factors including driving style.

For example, a 50 kWh battery in an EV can be expected to provide a certain range, and this information can be used to compare it with a similar EV that has a 75 kWh battery. This helps us make informed decisions about the range and energy efficiency of different EVs. Additionally, using kilowatt-hours to measure the energy storage capacity of EV batteries in the same way that electricity is sold and billed makes it a convenient and widely recognised unit of measurement.

The actual amount of energy that can be used from the battery will depend on a variety of factors, including the discharge rate (how fast energy is used), the temperature and the battery's state of health. Battery capacity can degrade over time, especially if the battery is frequently fast-charged and discharged or if it is exposed to extreme temperatures.

An example of one of the latest high-voltage battery systems (Figures 5.2 and 5.3), is in the new Audi Q8 e-tron. This car operates with a nominal voltage of 396 V. Two different battery sizes are available:

▶ Q8 50 e-tron – a usable capacity of 89 kWh
▶ Q8 55 e-tron and SQ8 e-tron – a usuable capacity of 106 kWh

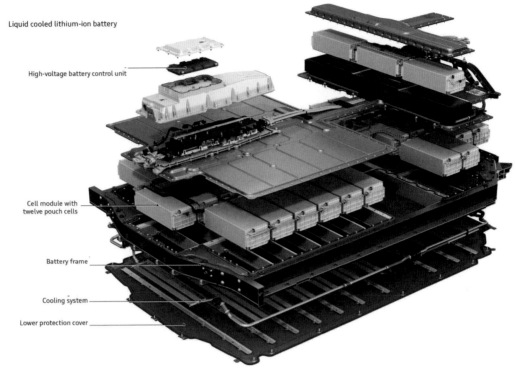

Liquid cooled lithium-ion battery

High-voltage battery control unit

Cell module with
twelve pouch cells

Battery frame

Cooling system

Lower protection cover

Figure 5.2 Audi Q8 e-tron battery pack (Source: Audi Media)

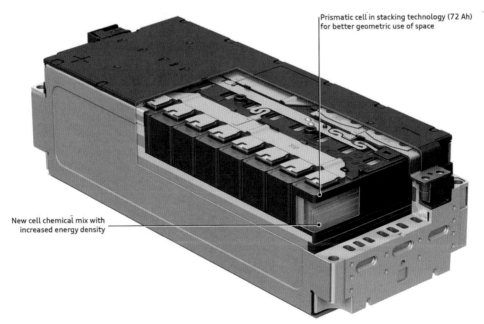

Prismatic cell in stacking technology (72 Ah)
for better geometric use of space

New cell chemical mix with
increased energy density

Figure 5.3 Battery module with 12 prismatic cells (Source: Audi Media)

107

In both cases, the space required for the drive battery is the same; thanks to further developments in cell technology and structure as well as in cell chemistry, the individual cells boast an increased energy density.

The prismatic cells used in battery production are assembled via a process called stacking technology, whereby the cell material is stacked in layers, thus filling the rectangular space much more efficiently – up to 20% more active cell material for power storage fit in the battery cell.

5.1.3 Battery life and recycling

Manufacturers usually consider the end of life for a battery to be when the battery capacity drops to 80% of its rated capacity. This means that if the original battery has a range of 100 km from a full charge, after 8–10 years of use, the range may reduce to 80 km. However, batteries can still deliver usable power below 80% charge capacity. A number of vehicle manufacturers have designed the battery to last the lifetime of the car.

Key Fact
Manufacturers usually consider the end of life for a battery to be when the battery capacity drops to 80% of its rated capacity.

The main sources of lithium for EV batteries are salt lakes and salt pans, which produce the soluble salt lithium chloride. The main producers of lithium are South America (Chile, Argentina and Bolivia), Australia, Canada and China. Lithium can also be found in rocks and mined – some are being opened in the UK.

Lithium can even be extracted from sea water. It is expected that recycling will become a major source of lithium. Worldwide reserves are estimated to be about 30 million tons. Around 0.3 kg of lithium is required per kWh of battery storage. Easy to extract resources may become a challenge, but, whilst opinions vary, many agree reserves will last over a thousand years. Nonetheless, recycling will become an important part of the lithium cycle when the process becomes economically viable (it is very close now).

The volume of lithium recycling at the time of writing is relatively small, but it is growing. Lithium-ion cells are considered nonhazardous, and they contain useful elements that can be recycled. Lithium, metals (copper, aluminium, steel), plastic, cobalt and lithium salts can all be recovered.

Safety First
Lithium-ion cells are considered nonhazardous, and they contain useful elements that can be recycled.

Lithium-ion batteries have a lower environmental impact than other battery technologies, including lead-acid, nickel-cadmium and nickel–metal hydride. This is because the cells are composed of more environmentally benign materials. They do not contain heavy metals (cadmium, for example) or compounds that are considered toxic, such as lead or nickel. Lithium iron phosphate is essentially a fertilizer. As more recycled materials are used, the overall environmental impact will be reduced.

All battery suppliers must comply with *The Waste Batteries and Accumulators Regulations 2009*. This is a mandatory requirement that means manufacturers take batteries back from customers to be reused, recycled or disposed of in an appropriate way.

At the time of writing, battery cells make up about 25% the price of a pure electric vehicle (2023). A problem to solve is growth in demand (in the EU) for raw materials, which is anticipated to be 25 times greater in 2030 than in 2015. One way forward is the 5R solution as shown in Figure 5.4.

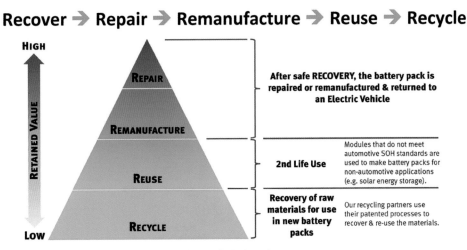

Figure 5.4 5R solution (Source: Autocraft EV Solutions)

Cells age differently, so only a proportion will be unusable at the end of a battery pack life. These cells (or modules) can be sorted and repurposed. Through cell replacement (5% to 30% of the cells), EV battery packs can theoretically be remanufactured to a state of health (SOH) of almost 100% multiple times.

> **Key Fact**
> Cells age differently, so only a proportion will be unusable at the end of a battery pack life.

The Autocraft EV battery system identifies, grades and produces different grades of packs:

▶ Grade A packs (Repair) for use in vehicles within original new spec
▶ Grade B packs (Remanufacture) for use in vehicles to a lower capacity spec
▶ Grade C packs (Reuse) for use in alternative markets
▶ Grade D packs (Recycle) made safe for material recycling partners

5.1.4 State of charge

State of charge (SOC) is a measure of the amount of energy left in a battery compared with the energy it had when it was full. It gives the driver an indication of how much longer a battery will continue to work before it needs recharging. It is considered as a measure of the short-term capability of a battery.

> **Definition**
> State of charge (SOC) is a measure of the amount of energy left in a battery compared with the energy it had when it was full.

However, it is more difficult to define SOC than it would first appear. It is defined as the available capacity expressed as a percentage of a given reference. This could be its

▶ rated capacity (as if new)
▶ latest charge-discharge capacity

Particularly on an EV, this can be a problem.

A vehicle that has a range of, say, 100 km on a fully charged, brand-new battery could reasonably expect a range of 50 km if it was 50% charged. However, after several years, the capacity of the battery when fully charged may only be 80% of what it used to be. An indication of 50% charge would now only give a 40 km range.

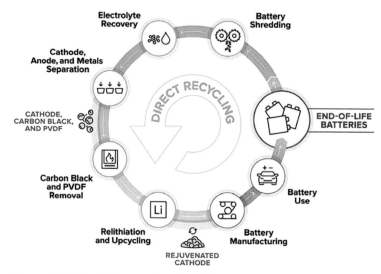

Figure 5.5 The EV battery 'circle of life' (Source: Argonne National Laboratory's ReCell R&D centre)

Because electric vehicles use the SOC to determine range, it should ideally be an absolute value based on the capacity of the battery when new. Several methods of estimating the state of charge of a battery have been used. Some are specific to particular cell chemistries. Most depend on measuring a parameter which varies with the state of charge.

The easiest way to monitor SOC is a voltage measurement, but this does depend on several factors. An open-circuit voltage will be higher than when current is flowing due to cell internal resistance. Temperature also has a big effect. Lithium-ion batteries also have a cell voltage that doesn't change that much between fully charged and fully discharged. Most are also actually operated between 80% and 20% as this reduces degradation over time. The voltage changes are, therefore, even smaller. Nonetheless, taking all factors into account, a voltage measurement under a known load gives a reasonable estimate of SOC.

It is also possible to calculate state of charge by measuring current and time (in or out). Current multiplied by time gives a suitable

value for SOC. Unfortunately, there are a few problems with this:

▶ The discharge current changes nonlinearly as the battery becomes discharged.
▶ To know how much charge it contains, the battery must be discharged.
▶ There are losses during the charge/ discharge cycle.

A battery will always deliver less during discharge than was put into it during charging. This is sometimes described as the coulombic efficiency of the battery. Temperature is once again an issue. However, if all factors are considered, a reasonable figure for state of charge can be calculated. Most battery manufacturers use coulombs in and coulombs out as a benchmark for their warranties.

5.1.5 State of health

The state of health (SOH) of a battery is a measurement that indicates the general condition of a battery and its ability to perform compared with a new battery. It considers charge acceptance, internal resistance, voltage and self-discharge. It is considered as a measure of the long-term capability of a battery.

Definition

The state of health (SOH) of a battery is a measurement that indicates the general condition of a battery and its ability to perform compared with a new battery.

SOH is an indication not an absolute measurement. During the lifetime of a battery, its performance deteriorates due to physical and chemical changes. Unfortunately, there is no agreed definition of SOH.

Cell impedance or cell conductivity is often used as a reasonable estimate of SOH. More complex systems monitor other parameters and involve a range of calculations. Because SOH is a figure relative to the condition of a new battery, the measurement system must collect and save data over time and monitor the change.

Counting the charge/discharge cycles of the battery is a measure of battery usage and can be used to indicate SOH, if compared to the expected values over time. This is because the capacity of lithium-ion cell deteriorates quite linearly with age or cycle life. The remaining cycle life can then be used as a measure of SOH.

At the time of writing, there is much discussion and related development of regulations that will force manufacturers to display or provide easy access to their vehicle's battery SOH. This is a challenge, however, because the SOH data obtainable from the battery management system (BMS) can be manufacturer-specific and not always accurate – so a level of accuracy will be required in the regulations too.

5.2 Types of battery

5.2.1 Introduction

As an introduction to batteries, it is interesting to compare their power densities with petrol. Power density means how much power is stored in a certain volume (litres, for example) or mass in kilograms. The watt-hour is a description of the power and is explained in more detail later.

Petrol is about 9,500 Wh/L and 12,800 Wh/kg. But engine efficiency is only around 25%, so this results in about 2,400 Wh/L and 3,200 Wh/kg.

Currently 700 Wh/L or 250 Wh/kg is achievable in a battery cell, but in the future 1400 Wh/L or 500 Wh/kg will be possible.

Batteries use a number of different chemistries that are defined by their cathode/anode materials. Here are some examples:

▶ lead acid
▶ nickel-cadmium
▶ nickel–metal hydride
▶ lithium-ion (graphite-transition metal oxide)
▶ lithium metal (lithium-transition metal oxide)
▶ lithium-sulphur
▶ metal-air

5.2.2 Lead-acid batteries (Pb–PbO$_2$)

Even after about 160 years of development and much promising research into other techniques of energy storage, the lead-acid battery is still the best choice for low-voltage

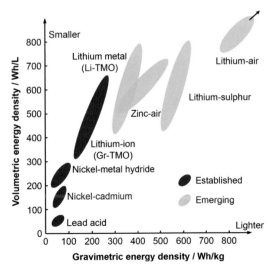

Figure 5.6 Comparing energy density of different chemistries

111

motor vehicle use. This is particularly so when cost and energy density are considered.

Key Fact

Gaston Planté was the French physicist who invented the lead-acid battery in 1859.

Incremental changes over the years have made the sealed and maintenance-free battery now in common use very reliable and long-lasting. This may not always appear to be the case to some end users, but note that quality is often related to the price the customer pays. Many bottom-of-the-range cheap batteries, with a 12-month guarantee, will last for 13 months!

The basic construction of a nominal 12 V lead-acid battery consists of six cells connected in series. Each cell, producing about 2 V, is housed in an individual compartment within a polypropylene, or similar, case.

Figure 5.7 depicts a cut-away battery showing the main component parts. The active material is held in grids or baskets to form the positive and negative plates. Separators made from a microporous plastic insulate these plates from each other.

However, even modern batteries described as sealed do still have a small vent to stop the pressure buildup due to the very small amount of gassing. A further requirement of sealed batteries is accurate control of charging voltage.

During charging, the electrolyte (sulfuric acid) becomes stronger and weaker during discharge. This means that measuring its relative density (specific gravity) is a good indicator of its state of charge.

5.2.3 Alkaline (Ni-Cad, Ni-Fe and Ni–MH)

The main components of the nickel-cadmium (Ni-Cad or NiCad) cell for vehicle use are as follows:

▶ positive plate – nickel hydrate (NiOOH)
▶ negative plate – cadmium (Cd)
▶ electrolyte – potassium hydroxide (KOH) and water (H_2O)

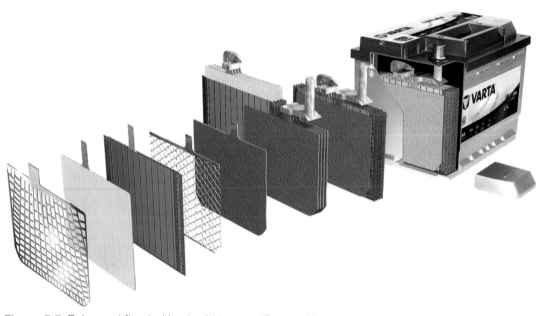

Figure 5.7 Enhanced flooded lead-acid battery (Source: Varta)

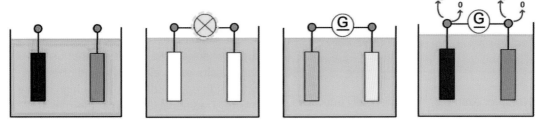

Figure 5.8 Battery discharge and charging process (left to right): Fully charged; discharging; charging; charging and gassing

Figure 5.9 Typical vehicle battery (Source: Bosch Media)

Key Fact

NiCad batteries do not suffer from overcharging because once the cadmium oxide has changed to cadmium, no further reaction can take place.

Nickel–metal hydride (Ni–MH or NiMH) batteries are used by some electric vehicles and have proved to be very effective. Toyota in particular has developed these batteries. The components of NiMH batteries include a cathode of nickel-hydroxide, an anode of hydrogen absorbing alloys and a potassium hydroxide (KOH) electrolyte. The energy density of NiMH is more than double that of a lead-acid battery but less than lithium-ion batteries.

The process of charging involves the oxygen moving from the negative plate to the positive plate and the reverse when discharging.

When fully charged, the negative plate becomes pure cadmium and the positive plate becomes nickel hydrate.

Water is lost as hydrogen (H) and oxygen (O_2) because gassing takes place all the time during charge. It is this use of water by the cells that indicates they are operating. The electrolyte does not change during the reaction. This means that a relative density reading will not indicate the state of charge.

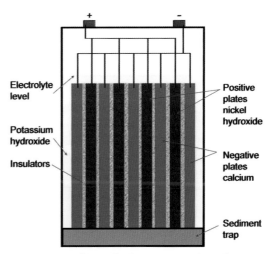

Figure 5.10 Simplified representation of a NiCad alkaline battery cell

113

5 Batteries

Key Fact

NiMH batteries are used by some electric vehicles and have proved to be very effective.

Toyota, for example, has continually improved its NiMH batteries by reducing size, improving power density, lowering weight, improving the battery pack/case and lowering costs. The current NiMH battery, which powers the third-generation Toyota Prius, costs 25% that of the battery used in the first generation.

Nickel-metal batteries are ideal for mass producing affordable conventional hybrid vehicles because of their low cost, high reliability and high durability. There are first-generation Prius batteries still on the road with over 200,000 miles and counting. That is why NiMH remains the battery of choice for some hybrid cars.

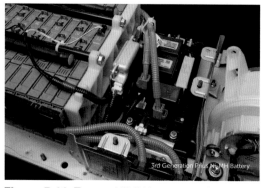

Figure 5.11 Toyota NiMH battery and management components (Source: Toyota)

Figure 5.12 Third-generation NiMH battery (Source: Toyota)

5.2.4 Sodium–nickel chloride (Na–NiCl$_2$)

Molten salt batteries (including liquid metal batteries) are a class of battery that uses molten salts as an electrolyte and offers both a high energy density and a high power density. Traditional 'use-once' thermal batteries can be stored in their solid state at room temperature for long periods of time before being activated by heating. Rechargeable liquid metal batteries are used for electric vehicles and potentially also for grid energy storage to balance out intermittent renewable power sources, such as solar panels and wind turbines.

Thermal batteries use an electrolyte that is solid and inactive at normal ambient temperatures. They can be stored indefinitely (over 50 years) yet provide full power in an instant when required. Once activated, they provide a burst of high power for a short period (a few tens of seconds) to 60 minutes or more, with output ranging from a few watts to several kilowatts. The high power capability is due to the very high ionic conductivity of the molten salt, which is three orders of magnitude (or more) greater than that of the sulfuric acid in a lead-acid car battery.

Key Fact

Thermal batteries use an electrolyte that is solid and inactive at normal ambient temperatures.

There has been significant development relating to rechargeable batteries using sodium (Na) for the negative electrodes. Sodium is attractive because of its high potential of 2.71 V, low weight, nontoxic nature, relative abundance, ready availability and low cost. In order to construct practical batteries, the sodium must be used in liquid form. The melting point of sodium is 98 °C (208 °F). This means that sodium-based batteries must operate at high temperatures between 400

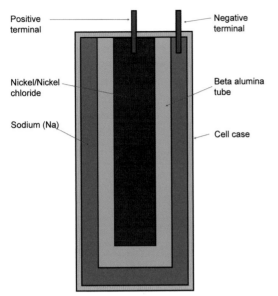

Figure 5.13 Sodium–nickel-chloride battery

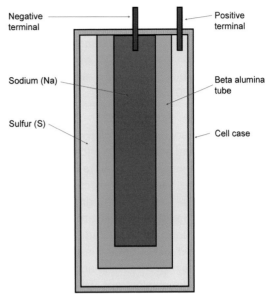

Figure 5.14 Sodium-sulfur battery

and 700 °C, with newer designs running at temperatures between 245 and 350 °C.

> **Safety First**
> Sodium-based batteries must operate at high temperatures between 400 and 700 °C, with newer designs running at temperatures between 245 and 350 °C.

5.2.5 Sodium-sulphur (Na-S)

The sodium-sulphur or Na-S battery consists of a cathode of liquid sodium into which is placed a current collector. This is a solid electrode of alumina (a form of aluminium oxide). A metal can that is in contact with the anode (a sulphur electrode) surrounds the whole assembly. The major problem with this system is that the running temperature needs to be 300–350 °C.

A heater rated at a few hundred watts forms part of the charging circuit. This maintains the battery temperature when the vehicle is not running. Battery temperature is maintained when in use due to current flowing through the resistance of the battery (often described as I^2R power loss).

Each cell of this battery is very small, using only about 15 g of sodium. This is a safety feature because, if the cell is damaged, the sulphur on the outside will cause the potentially dangerous sodium to be converted into polysulfides, which are comparatively harmless. Small cells also have the advantage that they can be distributed around the car. The capacity of each cell is about 10 Ah. These cells fail in an open-circuit condition and, hence, this must be considered, as the whole string of cells used to create the required voltage would be rendered inoperative. The output voltage of each cell is about 2 V.

5.2.6 Lithium-ion (Li-ion)

Lithium-ion technology is becoming the battery technology of choice, and it still has plenty of potential to offer. Today's batteries have an energy density of up to 140 Wh/kg or more in some cases but have the potential to go as high as 280 Wh/kg. Much research in cell optimisation is taking place to create a battery with a higher energy density and increased range. Lithium-ion technology is currently considered the safest.

115

Key Fact

Today's batteries have an energy density of approximately 140 Wh/kg or more in some cases but have the potential to go as high as 280 Wh/kg.

The Li-ion battery works as follows. A negative pole (anode) and a positive pole (cathode) are part of the individual cells of a lithium-ion battery together with the electrolyte and a separator. The anode is a graphite structure and the cathode is layered metal oxide. Lithium-ions are deposited between these layers. When the battery is charging, the lithium-ions move from the anode to the cathode and take on electrons. The number of ions, therefore, determines the energy density. When the battery is discharging, the lithium-ions release the electrons to the anode and move back to the cathode.

Key Fact

Inside a battery the anode and cathode are the opposite to the terminals outside the battery.

The electrode of a battery that releases electrons during discharge is called the anode, and the electrode that absorbs the electrons is the cathode. Internally, the battery anode is always negative and the cathode positive.

Useful work is performed when electrons flow through a closed external circuit. The following equations show one example of the chemistry, in units of moles, making it possible to use coefficient X.

The cathode (marked +) half-reaction is:

$$Li_{1-x}CoO_2 + xLi^+ + xe^- \rightleftarrows LiCoO_2.$$

The anode (marked -) half-reaction is:

$$xLiC_6 \rightleftarrows xLi^+ + xe^- + xC_6.$$

One issue with this type of battery is that in cold conditions, the lithium-ions' movement is slower during the charging process. This tends to make them reach the electrons on the surface of the anode rather than inside it. Also, using a charging current that is too high creates elemental lithium. This can be deposited on top of the anode covering the surface, which can seal the passage. This is known as lithium plating. Research is ongoing, and one possible solution could be to warm up the battery before charging.

Key Fact

Lithium-ion movement is slower during the charging process if the battery is cold.

Bosch is working on post-lithium-ion batteries, such as those made using lithium-sulphur technology, which promises greater energy density and capacity. The company estimates that the earliest the lithium-sulphur battery will be ready for series production is the middle of the 2020s.

There are several ways to improve battery performance. For example, the material used for the anode and cathode plays a major role in the cell chemistry. Most of today's cathodes consist of nickel-cobalt-manganese (NCM) and nickel-carboxyanhydrides (NCA), whereas anodes are made of graphite, soft or hard carbon or silicon carbon.

High-voltage electrolytes can further boost battery performance, raising the voltage within the cell from 4.5 to 5 volts. The technical challenge lies in guaranteeing safety and longevity while improving performance. Sophisticated battery management can further increase the range of a car by up to 10%, without altering the cell chemistry.

Key Fact

Sophisticated battery management can further increase the range of a car by up to 10%, without altering the cell chemistry.

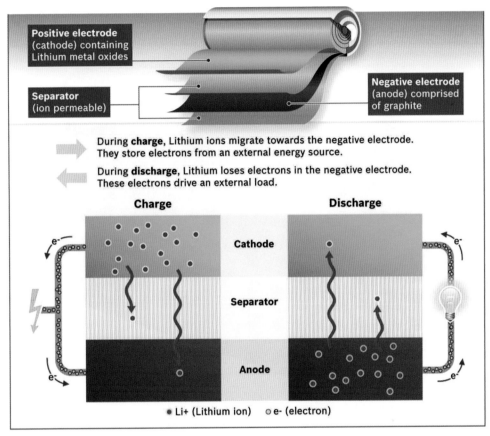

Figure 5.15 Basic operation of a lithium-ion battery (Source: Bosch Media)

5.2.7 Lithium-ion blade-battery case studies

The blade battery (Figure 5.17) is essentially a lithium–iron phosphate ($LiFePO_4$) battery that features a unique design resembling a blade. It inherits the high battery life characteristic of $LiFePO_4$ batteries, especially in winter conditions. The name "blade battery" is derived from its blade-like appearance.

The blade battery utilises high-safety $LiFePO_4$ technology and has achieved a 50% increase in volume and energy density compared to previous versions. It has undergone rigorous industry-standard tests, including the demanding acupuncture test. Electric vehicles equipped with blade batteries can achieve a range of over 600 km, showcasing their impressive performance.

Compared to conventional $LiFePO_4$ and ternary batteries (those containing three metallic elements), blade batteries excel in several aspects such as safety, volume utilisation, cycle life and structural strength. Although the blade battery is essentially a square, hard-shell battery, it adopts a long and thin design. The overall dimensions vary slightly across different models, with measurements typically around 960 × 90 × 13.5 mm. The thickness may differ, for instance, with the 138 Ah blade battery being approximately 12 mm thick and the 202 Ah blade battery measuring around 13.5 mm thick. The blade battery features threaded and

117

Audi Q8 55 e-tron quattro
Lithium-Ionen-Batterie mit 106 kWh netto (114 kWh brutto)
Lithium-ion battery with 106 kWh net (114 kWh gross)
11/22

Battery Junction Box (BJB)

Zellmodule mit zwölf prismatischen Zellen (72 Ah)
Cell module with twelve prismatic cells (72 Ah)

Gehäusedeckel
Housing cover

Batterie Management Controller (BMC)
Battery management controller (BMC)

Batterierahmen
Battery frame

Kühlsystem
Cooling system

Aluminium Fachwerkstruktur
Aluminium structure

Unterfahrschutz aus GFK
Lower protection cover made from GRP

Figure 5.16 Lithium-ion battery (Source: Audi Media)

Figure 5.17 Blade type cells – developments are ongoing (Source: Bosch Media)

The design of the lithium battery pack has undergone significant changes with the implementation of the cell-to-pack (CTP) module-free approach, which alters the cell structure. The conventional shell structure of the battery is eliminated, and the blade battery itself acts as both the beam and the battery cell. A honeycomb aluminium-plate design is adopted, with two high-strength aluminium plates affixed to the upper and lower sides, within which the blade batteries are arranged. This eliminates the need for traditional module connections, improves space utilisation and allows for the installation of more batteries in the same area.

During the assembly of the battery pack, the strength of the blade battery is utilised, and the battery pack is grouped to reduce the need for additional structural support. When subjected to random vibration loads, the module experiences significant deformation, while the blade battery

welded platform-type pole terminals, allowing for flexible connection between batteries during assembly. The negative-side terminal is equipped with an explosion-proof valve to enhance safety.

cell remains relatively unaffected. By modifying the battery pack structure and vertically inserting rectangular batteries into it, the design of the battery pack is simplified, and the space utilisation within the pack is improved.

The blade battery offers a flexible structure, allowing for customisation of the length of each battery to optimise space utilisation within the vehicle. Compared to ordinary lithium–iron phosphate batteries, blade batteries achieve a volume utilisation rate increase of over 50%, comparable to ternary lithium batteries.

Similar to the strength of a single chopstick compared to a bundle of chopsticks, the tight arrangement of hundreds of blade battery cells, each less than one meter in length, provides enhanced structural support to the vehicle's safety system in the event of a collision.

In terms of energy density, the range of blade batteries is close to that of ternary lithium batteries. Higher power storage capacity at the same volume or weight translates to higher volumetric or mass energy density, resulting in improved battery life for electric vehicles. The blade battery compensates for the lower energy density of ordinary lithium–iron phosphate batteries.

Regarding cycle life, lithium–manganese oxide batteries have the fewest number of cycles, ternary lithium batteries fall in the middle, and both lithium–iron phosphate batteries and blade batteries offer the highest number of cycles, aligning with the characteristics of lithium–iron phosphate chemistry.

The surface temperature of the blade battery typically ranges from 30 to 60 °C, and it is not prone to spontaneous ignition.

However, the blade battery exhibits relatively poor performance in low temperatures and is not resistant to cold conditions. This is a common drawback of lithium–iron phosphate batteries, and the battery's power output is significantly reduced in cold winter conditions.

The maintenance cost of the blade battery is higher compared to traditional batteries because it eliminates the need for a separate structural support system and relies on the battery cells themselves as support. As a result, ensuring the integrity of the battery becomes challenging, and if one battery cell is damaged, it can affect the rest of the batteries connected in series. The manufacturing cost and process complexity are also higher.

Blade batteries are manufactured using a laminated process rather than the traditional winding structure employed by conventional power batteries. The laminated structure provides more uniform current density and better internal heat dissipation, making it suitable for high-power charging and discharging. Consequently, blade batteries exhibit improved cycle characteristics and safety features. To address the lower energy density of lithium–iron phosphate, the blade battery optimises space utilisation through its arrangement pattern, similar to a box filled with ping-pong balls. While the individual density is low, the overall density is high, thereby improving space utilisation.

According to the manufacturer's data, the theoretical service life of the blade battery is 1.2 million km or 8 years, with a charge and discharge cycle life of over 3,000 times. In other words, if a car owner does not drive a total of 120,000 km within 8 years, the battery can still be used. However, its overall lifespan and power storage capacity will be reduced compared to its initial performance.

5.2.8 Tesla battery case study

Existing lithium-ion (Li-ion) battery cells are advanced technology, but even a small improvement translates as better range in an electric vehicle. Tesla recently patented a new battery cell. The 80 mm tall battery cell is described as 'tabless'. Tesla claims the cell will offer significant improvements over existing technologies that are shown in Figure 5.18.

Even if these figures are exaggerated a little, they are still a step change. Let's look at some of the technology and see why this could be a very important development. Most

Figure 5.18 Battery cell and expected improvements (Source: Tesla)

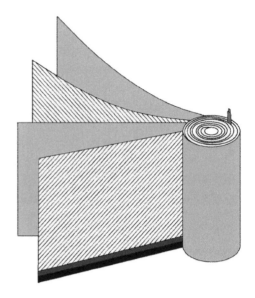

Figure 5.19 Battery cell construction (Source: Tesla patent application)

battery cells use a 'Swiss roll' design where the cathode, anode and separators are rolled together. The cathode and anode electrodes use a tab to connect to the positive and negative terminals of the cell cannister. Current flows through these tabs to connectors on the outside of the battery cell.

The resistance of the tabs is significant because they cause heat losses (known as I^2R losses), which increases with distance, If the current has to travel all the way along the cathode or anode to the tab and out of the cell, the losses start to add up. Also, reducing the number of tabs reduces costs and make manufacturing easier. The patent is called 'Cell with a Tabless Electrode'. Tesla wrote this in the patent application:

> The cell includes a first substrate having a first coating disposed thereon, wherein a second portion of the first substrate at a proximal end along the width of the first substrate comprises a conductive material. An inner separator is disposed over the first substrate. A second substrate is disposed over the inner separator. The second substrate has a second coating disposed thereon. The first substrate, the inner separator, and the second substrate in a successive manner, the first substrate, the inner separator, and the second substrate are rolled about a central axis.

The diameter of the new cell is double that of the existing ones used in Model 3 and Model Y vehicles. These currently use 2170 cells, which are produced by Panasonic at the Gigafactory in Nevada. Doubling the diameter of a cell increases the volume by four. Therefore, if the new volume is used efficiently, by reducing the tabs, for example, capacity is increased. Costs are also reduced because of fewer casings and cells per pack. Less than a 1,000 of these new cells could be used in a battery pack.

The relatively simple change, of using a larger cell with a tabless design, could result in significant improvements in overall performance. It is quite likely that material and chemical changes will also be made. This really could be a step change as Tesla suggests.

5.2.9 Fuel cells

The energy of oxidation of conventional fuels, which is usually manifested as heat, can be converted directly into electricity in a fuel cell. All oxidations involve a transfer of electrons between the fuel and oxidant, and this is employed in a fuel cell to convert the energy directly into electricity. All battery cells involve

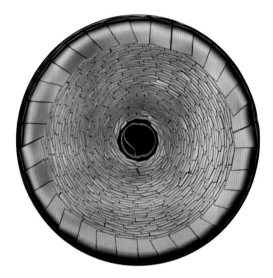

Figure 5.20 Battery cell (Source: Tesla)

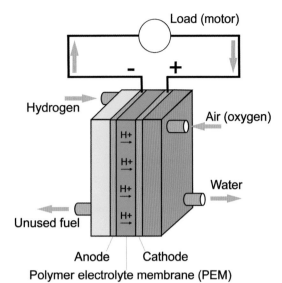

Figure 5.21 Proton exchange membrane fuel cell (polymer electrolyte membrane)

an oxide reduction at the positive pole and an oxidation at the negative during some part of their chemical process. To achieve the separation of these reactions in a fuel cell, an anode, a cathode and electrolyte are required. The electrolyte is fed directly with the fuel.

Key Fact
The energy of oxidation of conventional fuels can be converted directly into electricity in a fuel cell.

It has been found that a fuel of hydrogen when combined with oxygen proves to be a most efficient design. Fuel cells are very reliable and silent in operation but are quite expensive to construct.

Operation of a fuel cell is such that as hydrogen is passed over an electrode (the anode), which is coated with a catalyst, the hydrogen diffuses into the electrolyte. This causes electrons to be stripped off the hydrogen atoms. These electrons then pass through the external circuit. Negatively charged hydrogen anions (OH–) are formed at the electrode over which oxygen is passed such that it also diffuses into the solution.

These anions move through the electrolyte to the anode. Water is formed as the by-product of a reaction involving the hydrogen ions, electrons and oxygen atoms. If the heat generated by the fuel cell is used, an efficiency of over 80% is possible, together with a very good energy density figure. A unit consisting of many individual fuel cells is often referred to as a stack.

Key Fact
A unit consisting of many individual fuel cells is often referred to as a stack.

The working temperature of these cells varies, but about 200 °C is typical. High pressure is also used and can be of the order of 30 bar. A real challenge with a hydrogen car is that it is nowhere near as efficient as a pure battery EV (Figure 5.22).

Safety First
The working temperature of fuel cells varies but about 200 °C is typical. High pressure is also used and can be of the order of 30 bar.

Hydrogen **and electric drive**
Efficiency rates in comparison using eco-friendly energy

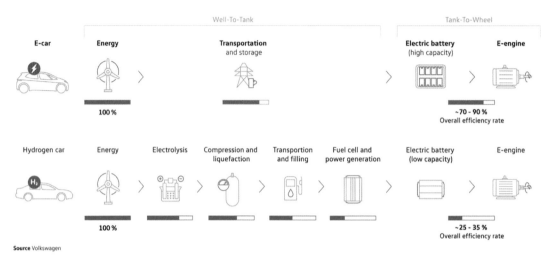

Source Volkswagen

Figure 5.22 Vehicle efficiencies compared (Source: Volkswagen)

5.2.10 **Super-capacitors**

Super- or ultra-capacitors are very high–capacity but (relatively) low-size capacitors. These two characteristics are achieved by employing several distinct electrode materials prepared using special processes. Some state-of-the-art ultra-capacitors are based on high surface area, ruthenium dioxide (RuO_2) and carbon electrodes. Ruthenium is extremely expensive and available only in very limited amounts.

> **Definition**
> Super- or ultra-capacitors are very high–capacity but (relatively) low-size capacitors.

Electrochemical capacitors are used for high power applications such as cellular electronics, power conditioning, industrial lasers, medical equipment and power electronics in conventional, electric and hybrid vehicles. In conventional vehicles, ultra-capacitors could be used to reduce the need for large alternators for meeting intermittent high-peak power demands related to power steering and braking. Ultra-capacitors recover braking energy dissipated as heat and can be used to reduce losses in electric power steering.

One system in use on a hybrid bus uses 30 ultra-capacitors to store 1600 kJ of electrical energy (20 farads at 400 V). The capacitor bank has a mass of 950 kg. Use of this technology allows recovery of energy when braking, which would otherwise have been lost because the capacitors can be charged in a very short space of time. The energy in the capacitors can also be used very quickly for rapid acceleration.

> **Key Fact**
> Capacitors can be charged in a very short space of time (compared with batteries).

5.2.11 **Comparing energy storage methods**

As a summary to this section, the following table compares the potential energy density of several types of battery. Wh/kg means watt-hours per

Table 5.2 Voltages and energy densities of batteries and storage devices (estimated for purposes of comparison)

Battery type	Specific energy (Wh/kg)	Energy density (Wh/l)	Specific power (W/kg)	Nominal cell voltage (V)	Amp-hour efficiency	Internal resistance (Ohms)	Operating temperature (°C)	Self-discharge (%)	Life cycles to 80%	Recharge time (h)
Lead-acid	20–35	54–95	250	2.1	80%	0.022	Ambient	2%	800	8 (1 hour to 80%)
Nickel–cadmium (Ni-Cad)	40–55	70–90	125	1.35	Good	0.06	-40 to +80	0.5%	1,200	1 (20 min to 60%)
Nickel–metal hydride (Ni-MH)	65	150	200	1.2	Quite good	0.06	Ambient	5%	1,000	1 (20 min to 60%)
Sodium–nickel chloride (Na-Ni-Cl)	100	150	150	2.5	Very high	Very low (increasing at low charge level)	300–350	10%/day	>1,000	8
Lithium-ion (Li-ion)	140	250–620	300–1,500	3.5	Very good	Very low	Ambient	10%/month	>1,000	30 min (to 80%)
Zinc-air	230	270	105	1.2	n/a	Medium	Ambient	High	>2,000	10 min
Aluminium-air	225	195	10	1.4	n/a	High hence low power	Ambient	>10%/day but if air removed very low	1,000	10 min
Sodium-sulphur	100	150	200	2	Very good	0.06	300–350	Quite low if kept warm	1,000	8
Hydrogen fuel cell	400		650	0.3–0.9 (1.23 open circuit)						n/a
Super-capacitor	1–10		1,000–10,000							Seconds
Flywheel	1–10		1,000–10,000							Seconds

kilogram. This is a measure of the power it will supply and for how long per kilogram.

5.3 Battery technologies

5.3.1 Ambient temperature

A study a few years ago by the American Automobile Association (AAA)[1] showed that the impact of temperature on EVs is significantly more than was expected. The organisation tested five models:

▶ BMW i3
▶ Chevrolet Bolt EV
▶ Nissan LEAF
▶ Tesla Model S
▶ Volkswagen e-Golf

Each car was operated at −7 °C and 35 °C. All had a similar response and, on average, lost about 12% range at the low temperature. This was a quite small loss but did not include the use of any cabin heating. When HVAC systems were activated, the range loss averaged 41%. This did not include seat or steering wheel heaters or headlights, which would impact the range further.

Focused on two EVs[2]:

▶ Tesla Model 3 with a 310-mile range rating
▶ Nissan LEAF with a 151-mile range rating

Testing was done on a track when the outside temperature averaged between −18 °C and −12 °C. The Tesla used up the equivalent of 121 miles to cover 64 actual miles, leaving a displayed range remaining of 189 miles. The Nissan used 141 miles of stated range to travel 64 miles, leaving just 10 miles showing on its range display.

Lithium-ion battery components develop increased resistance at low temperatures, and this limits how much power they can hold as well as how fast they can be charged or discharged.

5.3.2 Battery temperature

Electrical current flow in battery cells and the associated connections causes heat, so

Figure 5.23 Lithium-ion battery pack

cooling is, therefore, vital. This heating effect is proportional to the square of the current flowing multiplied by the internal resistance of the cells and connections (I^2R). Internal resistance of the cells rises when cold.

Many lithium battery cells should not be fast-charged when their temperature is below 0 °C. Battery management systems can handle this low temperature issue, but lithium cells also begin to degrade quickly if their temperature is too high (typically over 45 °C), and there are safety concerns with operation at high temperatures.

> **Key Fact**
> Many lithium battery cells should not be fast-charged when their temperature is below 0 °C.

The temperature of cells also needs to be kept constant across the pack. Uneven temperatures can lead to degraded performance and potential thermal events (Figure 5.24).

For efficient liquid cooling, there are two options:

▶ indirect cooling using a liquid that is pumped around the battery and passed through a cooler radiator or similar
▶ direct immersion cooling where the battery cell components are covered in a cooling agent

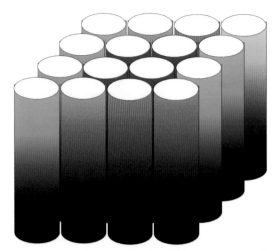

Figure 5.24 Representation of heat generated inside a battery pack (the penguin effect!)

Cooling the battery with a dielectric oil (the cooling agent) which is then pumped out to a heat exchanger system is an effective solution and offers good control. Figure 5.25 shows an example of this.

Currently, the most widely used cooling approach is the indirect cooling method using a traditional cooling agent. As the demand of energy volumetric density, safety and power density increases, there is an urgent need for a safer, more efficient cooling system design and cooling agents. Direct immersion cooling is the safe recommendation.

Direct liquid cooling/heating is more effective and takes up less volume provided that the heat transfer fluid is a safe and stable dielectric. Direct immersion cooling offers a safe, efficient, simplified design that enables more compact packaging.

> **Key Fact**
> Direct liquid cooling/heating is more effective and takes up less volume provided that the heat transfer fluid is a safe and stable dielectric.

Selection of the liquid for direct immersion cooling of electronics mustn't be made on the basis of heat transfer characteristics alone. Chemical compatibility of the coolant with the cells, electronics (control units) and other packaging materials must be a prime consideration to help avoid

▶ short circuits
▶ corrosion
▶ cross contamination
▶ flammability risk

There are several materials factors to be considered for fast charge of cells, because fast charging has four main limiting factors:

▶ lithium plating
▶ particle cracking
▶ atomic rearrangement
▶ temperature rise

These four main factors must be taken into consideration when determining the optimal fast charge profile. If this is not done, then significant reduction in cell capacity will be the result. Increasing the charging time by changing cell chemistry or design will have a cost to either lifetime or energy density.

Claimed charging times on some new battery designs vary from a few minutes to a few hours. The more ambitious fast charge times would have to overcome significant barriers to become a reality. As a bare minimum, a highly efficient thermal management system and detailed attention to component packaging is needed.

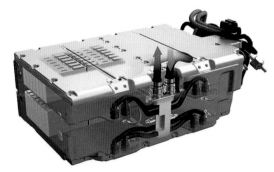

Figure 5.25 Battery pack with coolant flow connections (Source: Porsche Media)

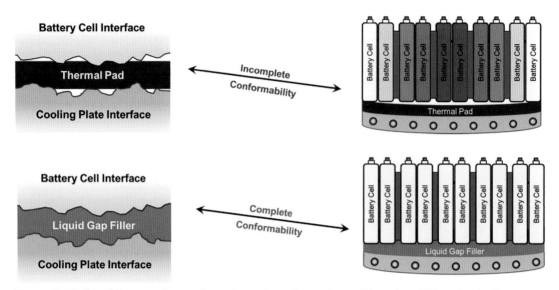

Figure 5.26 Gap fillers can have a huge impact on thermal conditions in a lithium-ion battery (Source: Solvay)

The use of dielectric fluid as a cooling medium enables the battery to have a higher power capability than the original design. This higher power capability allows for faster charging times, reducing the charge time from 2 hours to less than 30 minutes. This is particularly important as the expectations for electric vehicles have shifted towards larger batteries and higher performance, making faster charging essential.

The immersion cooling technology, which was tested on a Volvo XC60 plug-in hybrid by TotalEnergies and Ricardo, offers several advantages. It improves efficiency, safety and performance while also extending battery life and reducing costs and risks for vehicle manufacturers. Importantly, the immersion cooling system can be integrated into existing vehicle architectures without requiring significant modifications.

By utilising immersion cooling, the battery's thermal management is enhanced, resulting in faster charge times and improved overall performance. This advancement contributes to making electric vehicles more attractive to consumers by providing better performance, increased safety and reduced costs.

Furthermore, the immersion cooling technology has demonstrated its ability to prevent fires, highlighting its superior safety features compared to traditional cooling systems.

In addition to the performance and safety benefits, immersion cooling can also lead to weight and cost reductions for vehicles. Through a simpler design and the use of lighter materials, the battery cost can be reduced by approximately 6%, resulting in a lighter overall vehicle.

Overall, the use of immersion cooling technology in mass-produced vehicles is a significant development that offers advantages in terms of faster charging, improved performance, increased safety and reduced costs.

Figure 5.27 Battery pack (Source: Ricardo)

5.3.3 **Thermal runaway**

Lithium-ion batteries can, if overheated, go into a thermal runaway process. This can be separated into three stages (Figure 5.28).

An exothermic reaction increases the battery temperature and, therefore, the internal pressure of the Li-ion battery. Gas evolution also increases the pressure of the cell. If the cell is equipped with a pressure relief valve (not found on pouch cells), this valve will open and release flammable organic compounds. A pouch cell may burst if internal pressure is too high.

Emission of organic carbonates can be seen as white smoke. During further heating of the cell, the colour of the smoke turns into grey by emitting active electrode material (mainly graphite particles). This thermal runaway process heats the cell up to 700–1000 °C. This high temperature may affect adjacent cells and cause a chain reaction. With the organic solvent, the conducting salt, $LiPF_6$, is also emitted and reacts as follows:

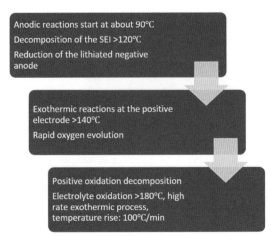

Figure 5.28 Thermal runaway stages. The SEI (solid electrolyte interphase) layer is a component of lithium-ion batteries formed from the decomposition of materials associated with the electrolyte of the battery

Safety First

An exothermic reaction increases the battery temperature and, therefore, the internal pressure of the Li-ion battery.

▶ When heated in dry environments, the salt decomposes.
▶ In contact with water/air moisture, toxic hydrogen fluoride (HF) gas is created.

If a thermal runaway occurs, a large number of different kinds of chemicals are generated. Combustion reactions mainly create

▶ CO, CO_2 from organic materials
▶ NOx, HF
▶ low molecular weight organic acids, aldehydes, ketones

Although a Li-ion battery fire should not normally be extinguished with pure water, using plenty of water may be reasonable because it cools the surrounding cells to avoid a following process. Additionally, many of the emitted particles and toxic gaseous compounds will bind and be diluted by water.

5.3.4 **Solid-state batteries**

Bosch and many others are working on a new battery technology for electric cars that could be production-ready in the near future. Solid-state cells could be a breakthrough technology as the batteries charge faster and are smaller than current technologies (figure 5.30).

With the new solid-state cells, there is the potential to more than double energy density and, at the same time, reduce the costs considerably further. A comparable electric car that has a driving range today of 150 km would be able to travel more than 300 km without recharging at a lower cost. Engineers are working on further refining the technology and, in doing so, making electromobility a more practical proposition. By 2025, the company forecasts that roughly 15% of all new cars built worldwide to have at least a hybrid powertrain.

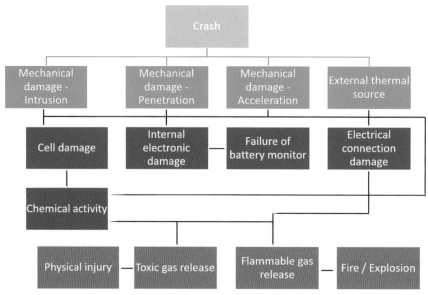

Figure 5.29 Battery harmful event flowchart

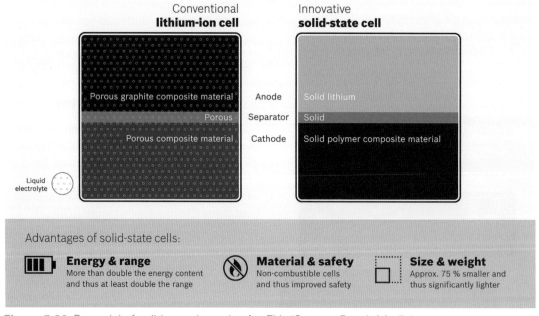

Figure 5.30 Potential of solid-state batteries for EVs (Source: Bosch Media)

The performance of an energy storage device can be improved by various methods. For example, in cell chemistry, the material that the positive and negative poles (cathode and anode) are made of plays a major role.

In current lithium-ion batteries, one of the reasons energy capacity is limited is because the anode consists, to a large degree, of graphite. Using solid-state technology means the anode can be made out of pure lithium, which considerably increases storage capacity. In addition, the new cells function without ionic liquid, which means they are not flammable.

5.3.5 Swarm intelligence

The older the batteries get, the lower their performance and capacity and the shorter the range of the vehicle. To help batteries last longer, Bosch is developing new cloud services that supplement the individual vehicle's battery management systems. Smart software functions in the cloud continually analyse the battery status and take appropriate action to prevent or slow down cell aging. These measures can reduce the wear and tear on the battery, the most expensive component of an electric vehicle, by as much as 20%.

> **Key Fact**
> The older the batteries get, the lower their performance and capacity and the shorter the range of the vehicle.

Real-time data gathered from the vehicle and its surroundings play a key role here. The cloud services use this data to optimise every single recharging process and to provide drivers with tailored driving tips on how to conserve battery power via the dash display. The aim is to optimise battery performance, thus benefiting both drivers and fleet operators within the ecosystem.

According to experts, the average service life of today's lithium-ion batteries is 8 to 10 years or between 500 and 1,000 charge cycles. Battery makers usually guarantee mileage of between 100,000 and 160,000 km.

However, rapid battery charging, high numbers of charge cycles, an overly sporty driving style and extremely high or low ambient temperatures are all sources of stress for batteries, which makes them age faster.

Bosch's cloud-based services are designed to recognise – and counter – these stress triggers. All battery-relevant data (e.g., current ambient temperature, charging habits) is first transmitted in real time to the cloud, where machine-learning algorithms evaluate the data. With these services, Bosch is not only offering a window into the battery's current status at all times but enabling a reliable forecast of a battery's remaining service life and performance to be made for the first time. Previously, it was not possible to make any accurate forecast of how quickly an electric vehicle battery would wear out.

The smart software functions use of the swarm principle: the algorithms used for analysis evaluate data gathered from an entire fleet not just from individual vehicles. Swarm intelligence is the key to identifying more of the stress factors for vehicle batteries and to identifying them more quickly.

The new insights gained into a battery's current status enable Bosch to actively protect it against aging. To give one example: Fully charged batteries age more quickly at particularly high or low ambient temperatures. Bosch's cloud services thus ensure that batteries are not charged to 100% when conditions are too hot or too cold. By reducing the battery charge by only a few percentage points, the battery is protected against inadvertent wear and tear.

Figure 5.31 Charging data is collected in the cloud (Source: Bosch Media)

Figure 5.32 A charging strategy based on the cloud data means optimum charging strategies can be used (Source: Bosch Media)

Data in the cloud will also help improve battery maintenance and repair. As soon as a battery fault or defect is identified, for example, the driver or fleet operator can be notified. This increases the chances that a battery can be repaired before it becomes irrevocably damaged or stops working altogether. Finally, the cloud services also optimise the recharging process itself.

> **Key Fact**
> Data in the cloud will also help improve battery maintenance and repair.

The recharging process is where there is a risk that the battery cells will permanently lose some of their performance and capacity. Smart software in the cloud can calculate an individual charge curve for each recharging process, regardless of whether it takes place at home or elsewhere. This means the battery is recharged to the optimum level, helping conserve the cells. Existing apps with charge timers merely allow drivers to time the recharging process so that it is carried out when demand for electricity is low. Optimising both fast and slow charging and controling electricity and voltage levels during the recharging process, prolongs battery life.

5.4 Research and development

5.4.1 Battery research

The increasing demand for battery raw materials, driven by the rising registrations of electrified vehicles, has raised concerns about the extraction and availability of these materials. While lithium and cobalt have been the subject of criticism for years due to ethical and environmental concerns, recent headlines have focused on the raw material nickel.

The desire for a secure supply, cost reduction and the ethical sourcing of materials has led to ongoing changes in the composition of cell chemistry in batteries. Currently, lithium-ion battery technology is predominantly used in electric cars, but this term encompasses a wide range of possible material combinations. The choice of battery raw materials has a significant impact on the storage capacity, safety, thermal stability and service life of the battery cell.

However, the challenge lies in finding the right balance between technical feasibility and accommodating overriding political factors when determining the battery composition. Adapting the composition of batteries to address political considerations is a complex task that requires careful evaluation of various factors, including performance, cost, availability and environmental and ethical concerns.

As the demand for electrified vehicles continues to grow, it is important for industry stakeholders to work towards sustainable and responsible sourcing of battery raw materials, explore alternative material options and invest in research and development to improve the efficiency and performance of batteries. Additionally, efforts to diversify supply chains and reduce dependence on specific regions can help mitigate potential disruptions caused by geopolitical events.

Lithium ions migration from the anode to the cathode during battery discharge and the reverse process during charging. The active materials in the electrodes play a crucial role in this ion insertion process. On the anode side, graphite is commonly used as the active material due to its ability to intercalate lithium ions. In some cases, silicon is added to graphite to enhance its performance. Both graphite and silicon are considered noncritical materials because they are widely available.

On the cathode side, there are various material compositions that can be used. Two common examples are nickel manganese cobalt (NMC) oxides and nickel cobalt aluminium (NCA) oxides. These cathode materials contain the rare element cobalt, which has raised concerns

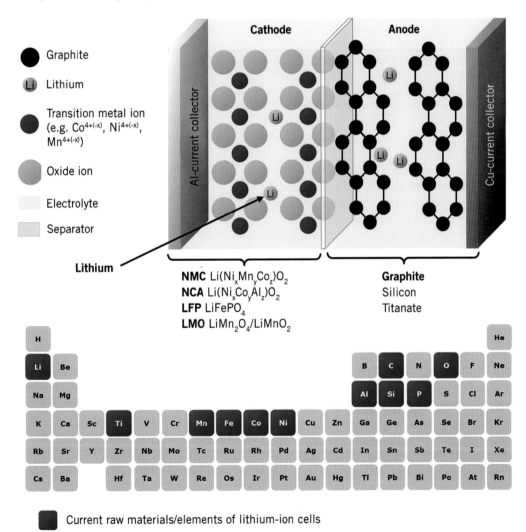

Figure 5.33 Battery cell elements[3]

due to its limited availability and ethical sourcing issues.

However, it's important to note that there are ongoing efforts to reduce or eliminate the use of cobalt in battery chemistries. One alternative is the lithium iron phosphate (LFP) cathode, which is cobalt-free. LFP batteries have gained popularity, particularly in certain applications such as electric buses, where safety, long cycle life and cost-effectiveness are important considerations.

The choice of cathode material in battery cells depends on several factors, including performance requirements, energy density, cost and availability of raw materials. The industry continues to explore new material combinations and chemistries to improve battery performance, reduce reliance on critical materials and address environmental and ethical concerns associated with raw material extraction.

Overall, the selection of battery cell raw materials is a complex process that involves balancing performance, availability, sustainability and cost considerations.

The use of lithium in batteries is not a coincidence but, rather, a result of its exceptional properties that make it well-suited for this application. Lithium is the lightest solid element, with a very low molar mass of 6.94 g/mol. This low mass contributes to the gravimetric energy density of batteries. Despite the fact that battery systems in electric vehicles (BEVs) weigh several hundred kilograms, the amount of lithium used typically ranges from 5 to 15 kg. This means that lithium can store a significant amount of energy relative to its weight.

Additionally, lithium's small ionic radius of 76 pm, derived from its position in the periodic table, is beneficial for the volumetric energy density of batteries. The compact size of lithium ions allows for efficient packing within the battery, increasing the amount of energy that can be stored in a given volume.

Furthermore, lithium has the most negative normal potential (-3.04 V) among all elements according to the electrochemical series. This means that lithium can achieve a high cell voltage, which is advantageous for battery performance. The high cell voltage contributes to the overall energy density of lithium batteries.

As a result of these characteristics, lithium enables batteries to achieve high energy densities of 250 Wh/kg and above. This high energy density is crucial for the efficient storage and delivery of electrical energy, making lithium an excellent choice for battery applications, particularly in the context of electric vehicles and portable electronics.

It's important to note that other factors, such as battery chemistry, cell design and manufacturing processes, also play significant roles in determining the overall performance of lithium batteries. Ongoing research and development efforts are focused on further improving the energy density, safety and cost-effectiveness of lithium-based battery technologies.

Researchers are exploring alternatives to lithium-based batteries to reduce dependence on lithium and improve cost-effectiveness. One such alternative is the sodium-ion battery (SIB). Sodium, being in the same group as lithium in the periodic table, shares some similarities and represents a plausible alternative. Sodium is more abundant and, therefore, offers a cost advantage compared to lithium. Additionally, certain components of the battery, such as the copper current collector and electrolyte, can be replaced with cheaper alternatives without compatibility issues.

However, it's important to note that sodium-ion technology also has significant drawbacks. The larger ionic radius of sodium leads to a reduced volumetric energy density compared to lithium, and it causes significant volume changes (known as breathing) during cell operation due to ion storage and removal at the electrodes. Furthermore, finding suitable materials for the anode in sodium-ion batteries is challenging, as graphite, commonly used in lithium-ion batteries, cannot be used.

While there are promising developments in lithium-free battery technologies, they are currently mostly research topics or niche products. It will take time for these alternatives to gain significant market share and overcome the dominance of lithium-ion technology. Even the solid-state battery, often considered as a potential successor to lithium-ion, is still based on lithium chemistry.

According to experts, lithium-ion technology is expected to continue dominating the market for many years to come. However, ongoing research and development efforts in both lithium-based and lithium-free battery technologies are focused on improving energy density, safety, cost-effectiveness and sustainability to meet the evolving needs of various applications.

5.4.2 Toyota 1500 km battery

Toyota is currently in the process of developing advanced battery technology that they claim will significantly enhance the range of future electric vehicles (EVs) to an impressive 932 miles (1,500 km).

As part of its forthcoming EV plans, the automaker has announced its commitment to achieving a ground-breaking improvement in battery performance within the next five years. By 2026, Toyota intends to introduce new battery technology that will offer an extended range of 1,000 km (621 miles) through enhanced energy density, weight reduction and improved vehicle aerodynamics. They also aim to reduce costs by 20% compared to the current model while achieving a quick charging time of 20 minutes or less for a 10–80% power charge.

Toyota is also working on the development of affordable batteries that will contribute to the wider adoption and expansion of battery electric vehicles (BEVs). One such innovation is the implementation of a bipolar structure battery, which has already been utilised in the Aqua and Crown hybrid vehicles and will now be applied to BEVs. This battery utilises lithium iron phosphate (LFP) and is anticipated to be available between 2026 and 2027. This low-cost battery is projected to offer a 20% increase

in range compared to their current model (approximately 375 miles), with a significant 40% reduction in cost and a recharge time of 30 minutes or less (10–80% charge).

However, the real game-changer on the horizon is the development of all-solid-state batteries, which have the potential to revolutionise the distance that Toyota vehicles can travel on a single charge. Overcoming the long-standing challenge of battery durability, Toyota has made a technological breakthrough and is actively exploring its implementation in conventional hybrid electric vehicles (HEVs) while expediting its application in BEVs. The company is currently focused on developing a mass production method, with plans to commercialise the technology between 2027 and 2028.

This all-solid-state battery technology is expected to deliver a 20% improvement in range compared to the previously announced 1,000 km range, along with an even faster charging time of 10 minutes or less (10–80% charge). As a result, Toyota's BEVs could potentially achieve a remarkable range of up to 1,200 km (745 miles).

In tandem with these advancements, Toyota is also working on a higher-level specification battery that aims to provide an extraordinary 50% increase in range, pushing the boundaries to a remarkable 1,500 km (932 miles).

These are great ambitions and Toyota does have a great history of innovation. But, as with all claims about amazing new battery technologies, we will have to wait and see.

5.4.3 Fire detection

The development of a new battery safety sensor by Metis Engineering aims to address the limitations of existing battery management systems (BMS) in detecting potential battery failures and preventing catastrophic incidents in electric vehicles. The Production Battery Safety Sensor is designed to monitor battery health beyond what a BMS can achieve.

The sensor, which is currently in the beta version, is capable of detecting cell venting within seconds, providing early detection of thermal runaway and allowing more time for vehicle evacuation. It collects data on various environmental parameters such as volatile organic compounds (VOCs), pressure change, humidity, dew point and optional accelerometer readings to record shock loads. By analysing this data, including cross-referencing with cell temperatures, the sensor can identify cell-venting events.

The sensor is integrated into the vehicle's electrical system using a CAN interface and is placed inside the battery pack to "sniff the air" for VOCs emitted during cell venting. It relays the collected data to a control unit, such as the vehicle's electronic control unit (ECU), which can alert the driver and initiate safety measures, such as cutting off the circuit to the battery pack to prevent thermal runaway.

While existing BMS systems already monitor battery temperature and voltage changes, they have limitations in detecting cell issues in a timely manner. Temperature sensors in BMS are typically placed at certain points within the pack, and if a cell is located far from the sensor, it may not detect temperature changes in time. Voltage fluctuations can also be challenging to interpret due to the influence of other cells in parallel.

The new battery safety sensor offers an additional layer of protection by providing more comprehensive monitoring of battery health and early detection of potential failures. It is currently being tested by several electric vehicle original equipment manufacturers (OEMs) and battery manufacturers.

Thermal runaway in battery cells is a complex process that can lead to catastrophic failures if not properly managed. The stages of thermal runaway involve temperature increase, pressure buildup, venting and the potential for a self-oxidising reaction that can result in a fire and further propagation of the failure.

The sensor addresses these risks by monitoring key parameters and providing early warnings. By detecting temperature increases, the sensor can identify when a battery cell is becoming overheated, which can be caused by factors like excessive workload, manufacturing defects or aging. The rise in temperature leads to pressure buildup inside the cell, and if the pressure exceeds a certain threshold, venting occurs.

To further enhance safety, the optional accelerometer in the sensor enables the monitoring of shock loads and impact duration that the battery pack may experience during a collision. This information can help assess the condition of the battery pack and inform decisions regarding servicing, repurposing for energy storage systems, recycling or insurance claims.

Another important aspect monitored by the sensor is the dew point within the battery pack. Cooling systems are typically used to prevent overheating, but if the cooling is too effective and lowers the temperature below the dew point, condensation can occur, leading to potential shorting and thermal incidents. By monitoring the dew point, the sensor can provide a warning before condensation settles on the battery terminals.

Overall, this battery safety sensor aims to detect early signs of thermal runaway, such as temperature increases, pressure changes, shock loads and dew point variations. By providing timely warnings, it helps mitigate

Figure 5.34 Battery safety sensor (Source: Metis)

the risks associated with battery failures and enables appropriate preventive actions to ensure the safety and longevity of the battery pack.

5.4.4 Leading edge (lithium-air)

Scientists and engineers at Argonne, a technology research company, have announced a ground-breaking achievement in battery technology (Figure 5.35). They claim to have successfully developed a revolutionary battery with an energy density four times higher than traditional lithium-ion batteries (in the lab). This remarkable advancement means that electric vehicles (EVs) could potentially travel over a thousand miles (1600 km) on a single charge, and in the future, the battery could also power domestic planes and long-distance trucks.

The key innovation in this lithium-air battery lies in the use of a solid electrolyte instead of the conventional liquid electrolyte. The researchers assert that the lithium-air battery exhibits the highest projected energy density among all battery technologies currently being explored for the next generation beyond lithium-ion. Previous lithium-air designs, where lithium from a lithium metal anode moved through a liquid electrolyte to combine with oxygen during discharge, resulted in the formation of lithium peroxide (Li_2O_2) or superoxide (LiO_2) at the cathode.

This solid ceramic polymer electrolyte (CPE) is composed of a material made from cost-effective nanoparticle elements. Its introduction enables chemical reactions that generate lithium oxide (Li_2O) upon discharge. The chemical process involving lithium superoxide or peroxide only requires the storage of one or two electrons per oxygen molecule, whereas lithium oxide necessitates the storage of four electrons.

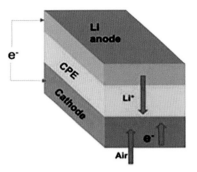

Figure 5.35 Lithium-air battery (Source: Argonne)

Through further refinement and development, Argonne anticipates that the lithium-air battery could achieve an unprecedented energy density of 1200 Wh/kg, four times better than existing lithium-ion batteries. To provide context, Tesla's latest 4680 cells, known for their high energy density, range between 272–296 Wh/kg.

5.4.5 Obituary

Battery science lost a giant in June 2023. Just at the point of completing the script for this book we heard the news that Nobel prize–winning scientist Professor John Goodenough (born in 1922) had died. He is credited as the father of the lithium-ion battery. Our respect and admiration. Learn more about his achievements at https://en.wikipedia.org/wiki/John_B._Goodenough

Notes

1 AAA: (the American Automobile Association, Inc.) https://www.aaa.com.4 Consumers Union: https://www.consumersinternational.org
2 Another study, conducted by Consumers Union.
3 Schmalz, M., Wagner, A., Lichtblau, F. et al. Range, Safety, Service Life - Battery Raw Materials and Their Technical Relevance. ATZ Electron Worldw 17, 8–13 (2022).

CHAPTER 6

Motors and control systems

6.1 Introduction

6.1.1 Types of motor

There are several choices for the type of drive motor, the first being between an AC or DC motor. The AC motor offers many control advantages but requires the DC produced by the batteries to be converted using an inverter. In much earlier systems, a DC shunt-wound motor rated at about 50 kW was a popular choice for smaller vehicles, but AC motors are now the most popular. Actually, the distinction is blurred. The drive motors can be classed as AC or DC, but it becomes difficult to describe the distinctions between an AC motor and a brushless DC motor.

> **Key Fact**
> The AC motor offers many control advantages but requires the DC produced by the batteries to be converted using an inverter.

6.1.2 Trends

The three-phase AC motor with a permanent magnet rotor is the choice for most manufacturers. This is because of its efficiency, size, ease of control and torque characteristics. It is powered by DC 'pulses'. This type is sometimes described as an electronically commutated motor (ECM).

6.2 Construction and function of electric motors

6.2.1 AC motors: basic principle

In general, all AC motors work on the same principle. A three-phase winding is distributed around a laminated stator and sets up a rotating magnetic field that the rotor follows. The general term is an AC induction motor. The speed of this rotating field and, hence, the rotor can be calculated:

$$n = 60\frac{f}{p}$$

where n is speed in rpm, f is frequency of the supply and p is the number of pole pairs.

6.2.2 Asynchronous motor

The asynchronous motor is often used with a squirrel-cage rotor made up of a number of pole pairs. The stator is usually three-phase and can be star or delta wound. The rotating

DOI: 10.1201/9781003431732-6

magnetic field in the stator induces an EMF in the rotor, which, because it is a complete circuit, causes current to flow. This creates magnetism, which reacts to the original field caused by the stator, and, hence, the rotor rotates. The amount of slip (difference in rotor and field speed) is about 5% when the motor is at its most efficient.

> **Key Fact**
>
> An asynchronous motor is often used with a squirrel-cage rotor.

6.2.3 Synchronous motor: permanent excitation

This motor has a wound rotor, known as the inductor. This winding is magnetised by a DC supply via two slip rings. The magnetism 'locks on' to the rotating magnetic field and produces a constant torque. If the speed is less than *n* (see previous section), fluctuating torque occurs and high current can flow. This motor needs special arrangements for starting

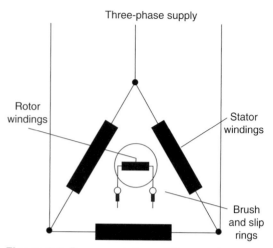

Figure 6.2 Representation of the synchronous motor

rotation. An advantage, however, is that it makes an ideal generator. The normal vehicle alternator is very similar.

> **Key Fact**
>
> A synchronous motor has a wound rotor known as the inductor.

6.2.4 DC motor: series wound

The DC motor is a well-proven device and was used for many years on electric vehicles such as milk floats and forklift trucks. Its main disadvantage is that the high current has to flow through the brushes and commutator.

The DC series-wound motor has well-known properties of high torque at low speeds. Figure 6.3 shows how a series-wound motor can be controlled using a thyristor and also provide simple regenerative braking.

> **Key Fact**
>
> The DC series-wound motor has well-known properties of high torque at low speeds and is ideal as a starter motor.

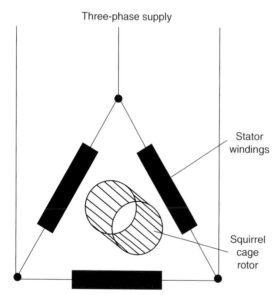

Figure 6.1 An asynchronous motor is used with a squirrel-cage rotor made up of a number of pole pairs

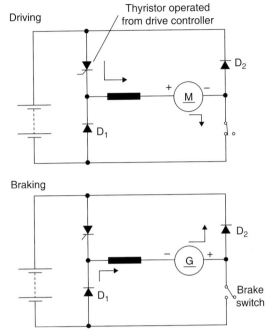

Figure 6.3 A series-wound motor can be controlled by using a thyristor and can also provide simple regenerative braking

6.2.5 DC motor: separately excited shunt wound

The fields on this motor can be controlled either by adding a resistance or using chopper control in order to vary the speed. Start-up torque can be a problem, but with a suitable controller, this can be overcome. The motor is also suitable for regenerative braking by increasing field strength at the appropriate time. Some EV drive systems only vary the field power for normal driving, and this can be a problem at slow speeds due to high current.

6.2.6 Electronically commutated motor

The electronically commutated motor (ECM) is, in effect, halfway between an AC and a DC motor. Figure 6.5 shows a representation of this system. Its principle is very similar to the synchronous motor, except the rotor contains permanent magnets and, hence,

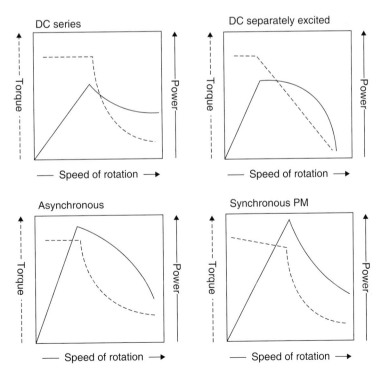

Figure 6.4 Torque characteristics

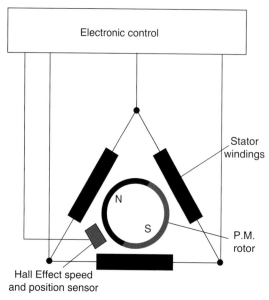

Figure 6.5 The EC motor is an AC motor

Labels in figure: Electronic control · Stator windings · P.M. rotor · Hall Effect speed and position sensor · N · S

---

no slip rings. It is sometimes known as a brushless motor. The rotor operates a sensor, which provides feedback to the control and power electronics. This control system produces a rotating field, the frequency of which determines motor speed. When used as a drive motor, a gearbox is needed to ensure sufficient speed of the motor is maintained because of its particular torque characteristics. Some schools of thought suggest that if the motor is supplied with square-wave pulses, it is DC and if supplied with sine-wave pulses, then it is AC.

> **Definition**
> ECM: electronically commutated motor.

These motors are also described as brushless DC motors (BLDC), and they are effectively AC motors because the current through them alternates. However, because the supply frequency is variable, it has to be derived from DC, and its speed/torque characteristics are similar to a brushed DC motor, it is sometimes called a DC motor.

> **Definition**
> BLDC: brushless DC motor.

It can also be called a self-synchronous AC motor, a variable-frequency synchronous motor, a permanent-magnet synchronous motor, or an electronically commutated motor (ECM) – I hope that's clear and prevents any further confusion! However, it is now the motor used for the majority of EVs.

The operating principle is shown in more detail in Figure 6.6. The rotor is a permanent magnet and the current flow through the coil determines the polarity of the stator. If switched in sequence and timed accordingly, the momentum of the rotor will keep it moving as the stator polarity is changed. Changing the switching timing can also make the rotor reverse. Overall, therefore, good control of the motor is possible.

> **Key Fact**
> The EC motor has a permanent magnet rotor.

The switching must be synchronised with the rotor position, and this is done by using sensors, Hall Effect, in many cases, to determine the rotor position and speed. If three coils or phases are used, as shown next, then finer control is possible as well as greater speed, smoother operation and increased torque. Torque reduces as speed increases because of back EMF. Maximum speed is limited to the point where the back EMF equals the supply voltage.

> **Key Fact**
> In any type of EC motor, supply switching must be synchronised with the rotor position. This is done by using position sensors.

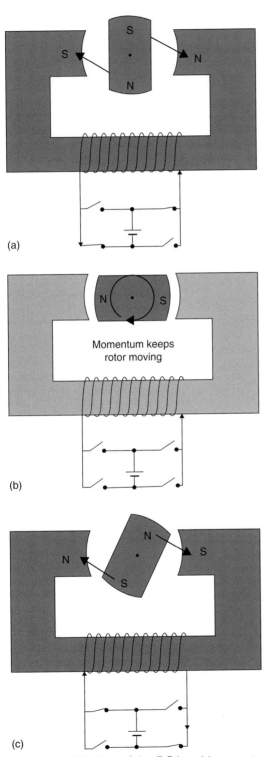

(a)

(b)

(c)

Figure 6.6 Principle of the DC brushless motor operation

Momentum keeps rotor moving

> **Definition**
>
> Hall Effect: The production of a voltage difference across an electrical conductor, transverse to an electric current in the conductor and a magnetic field perpendicular to the current. This effect was discovered by Edwin Hall in 1879 and is used for sensing rotation speed and position.

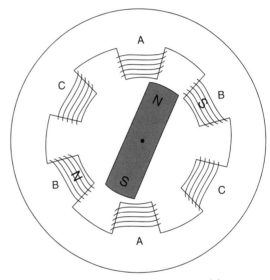

Figure 6.7 Three coils (three-phases) improve on the basic principle. Many more coils are used on real machines (see Figure 6.8)

Two typical motors are shown in Figures 6.8 and 6.9, one is integrated with the engine flywheel and the other a separate unit. Both are DC brushless motors and are water cooled.

6.2.7 Switched reluctance motor

The switched reluctance motor (SRM) is similar to the brushless DC motor. However, the major difference is that it does not use permanent magnets. The rotor is a form of soft iron and is attracted to the magnetised stator. The basic principle is shown in Figure 6.10 and an improved version in Figure 6.11.

Figure 6.8 Bosch integrated motor generator (IMG) also called integrated motor assist (IMA) by some manufacturers

Figure 6.9 Separate motor unit showing the coolant connections on the side and the three main electrical connections on top

Definition

SRM: switched reluctance motor.

Timing of the switching of the stator is again very important, but the key advantage is that no expensive rare earth magnets are needed. The raw materials for these are a source of political discussion, with China being the main supplier. Overall, the machine is very simple and, therefore, cheap. Early SRMs were noisy, but this has been solved by more accurate switching control.

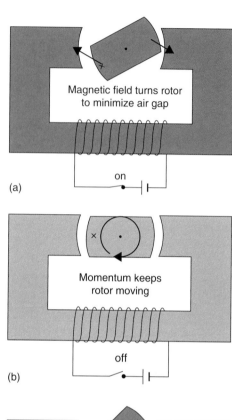

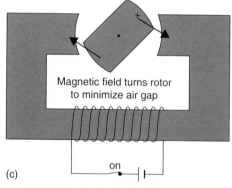

Figure 6.10 Basic principle of a switched reluctance motor

A company known as HEVT has developed a potentially game-changing alternative to induction and permanent magnet motors.

These motors have a much reduced cost volatility due to the use of zero rare earth minerals.

The company's patented switched reluctance motors (SRMs) provide high-performance

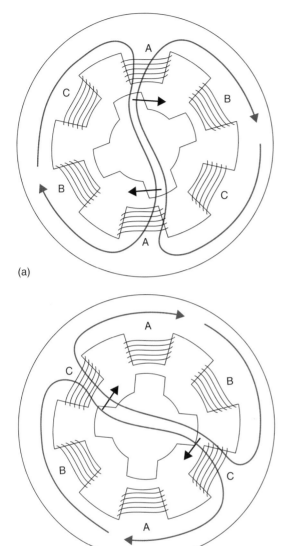

(a)

(b)

Figure 6.11 Improved SRM with additional poles (the rotor invariably has two fewer than the stator)

Figure 6.12 Switched reluctance motor stator (left) and rotor (right) without windings on the stator (Source: HEVT)

Key Fact

Switched reluctance motors do not work well as generators because of a nonmagnetic rotor.

The SRM control systems are similar to those for BLDC motors, but while they have slightly lower peak torque, their efficiency is maintained over a much wider speed and torque range. The SRM is effectively a powerful stepper motor.

6.2.8 **Motor efficiency**

Motor efficiency varies with type, size, number of poles, cooling and weight. The designers are always striving to get more out of smaller, lighter packages. In general, the efficiency of a BLDC ranges from about 80% for a 1 kW motor to 95% for a 90 kW motor.

Key Fact

In general, the efficiency of a BLDC ranges from about 80% for a 1 kW motor to 95% for a 90 kW motor.

alternatives to induction and permanent magnet motor/generators. One product is being rolled out to provide electrical assist to eBikes, but current motor technology ranges from approximately 150 W scalable to 1 MW, and it is expected they may appear in some EVs soon. However, they do not work well as generators because of the nonmagnetic rotor.

Efficiency is the ratio between the shaft output power and the electrical input power. Power output is measured in watts (W), so efficiency can be calculated:

$$\mathbf{p_{out}} / \mathbf{p_{in}}$$

143

where P_{out} = shaft power out (W) and
P_{in} = electric power into the motor (W).

The electrical power lost in the primary rotor and secondary stator winding resistance are also called copper losses. The copper loss varies with the load in proportion to the current squared and can be calculated:

$$I^2R$$

where R = resistance (Ω) and I = current (amp).

Definition

Motor efficiency is the ratio between the shaft output power and the electrical input power.

Other losses include

▶ iron losses: The result of magnetic energy dissipated when the motor's magnetic field is applied to the stator core.
▶ stray losses: The losses that remain after primary copper and secondary losses, iron losses and mechanical losses. The largest contribution to the stray losses is harmonic energies generated when the motor operates under load. These energies are dissipated as currents in the copper windings, harmonic flux components in the iron parts or leakage in the laminate core.
▶ mechanical losses: The friction in the motor bearings, in the fan for air cooling or a pump if water or oil cooling is used.

Figure 6.13 Motor efficiency

6.2.9 Motor torque and power characteristics

The torque and power characteristics of four types of drive motors are compared in Figure 6.14. The four graphs show torque and power as functions of rotational speed.

The characteristics torque and power curves of a typical ECM car drive motor compared to an internal combustion engine are shown as Figure 6.14.

6.2.10 Winding shapes and insulation

Using higher voltages (700 V+) means thicker insulation is needed on the motor windings. Solvent-based insulators, such as varnish, are not adequate for the job. However, careful design and the use of new materials can result in three key benefits:

▶ torque increase
▶ power increase
▶ size reduction

Innovative developments in the way stator windings are created and insulated are resulting in significant improvements in efficiency. Electric motor stator windings were traditionally round copper wires beneath a protective layer of enamel or varnish. Changing from round wire to rectangular helps fill the available space as represented in Figure 6.15.

When enamel is used as the insulator, it is very difficult to achieve the exact thickness required. Higher voltages require thicker insulation and/or insulation with a higher dielectric rating.

Conventional motors generally use enamelled wire with a film thickness of 0.1 mm or less. This enamel provides adequate insulation between wires in a phase, but because of the higher potential difference between phases, insulating paper is added.

It is not possible to increase the thickness of enamel coatings, so the enamelled magnet wire shown at the top in Figure 6.16 has been covered in extruded high-performance plastic (in this example Ketaspire© PEEK which is a polyetheretherketone resin). The wire at the bottom in Figure 6.16 is using PEEK and no enamel.

Typical Electric Motor Torque & Power Curves

Typical ICE Torque and Power Curves

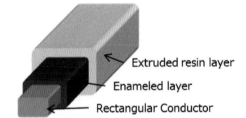

—Torque
—Power

Figure 6.14 ECM and ICE torque and power curves

Figure 6.15 Rectangular windings make more efficient use of the available space, even with similar or increased thickness of insulation

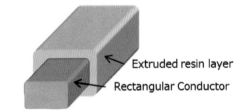

Figure 6.16 Enamel and/or extruded resin on a wire

This insulation can be applied in even thickness for predictable performance, and it has favourable adhesion to metals. Extrusion can be used in round, square and rectangle profiles, which enables motor designers to optimise the slot fill. Extrusion provides the added benefit of being a more efficient manufacturing process versus solvent-based coatings.

6.2.11 **Axial and radial windings**

In motors where the windings are made from separate segments, there are two main ways that the windings are positioned, axially as if wound around a shaft parallel to the axle, and radially as if wound around a radius of the central shaft (Figure 6.17).

A U.K. company YASA was founded on an innovative approach to the design and

manufacture of electric motors and controllers. In its motors, the windings consist of separate segments ideally suited to mass manufacture with minimal application engineering. Figure 6.18 shows a motor with axial windings.

The axial flux approach uses less materials such as copper, iron and permanent magnet than conventional motors, resulting in a significantly lower materials cost. The other key benefit is that these axial flux motors are smaller and lighter than any other motors in their class due to the more efficient use of key magnetic and structural materials. The

145

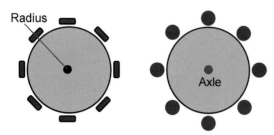

Figure 6.17 Radial (left) and axial (right) motor design

Figure 6.18 YASA axial electric motor, 260 Nm/180 kW 10 s peak, 65 mm length, 260 mm diameter, 12 kg, 15 W/kg (Source: YASA)

yokeless and segmented armature (YASA) motor topology also significantly reduces manufacturing complexity, making the motors ideally suited to automated volume production. The YASA 750, YASA P400 Series and YASA 400 axial flux motors achieve power densities

of up to 10 kW/kg and torque densities that are significantly better than the nearest competitive motor topology. The axial flux enables high power densities to be achieved at relatively low speeds (2000–9000 rpm), making the motors ideally suited to hybrid and generation applications.

6.2.12 Drive motor developments and case studies

Infinitum

Infinitum has made significant advancements in motor technology with their recent achievement of a 50% smaller and lighter axial-flux motor. This has been made possible by replacing the traditional iron core and copper windings with a printed circuit board (PCB) stator, which incorporates etched copper conductors. This innovative design eliminates core losses, such as torque ripple, cogging, stator hysteresis and eddy currents, resulting in a motor with higher overall efficiency.

In addition to its smaller and lighter form, the motor incorporates a liquid-cooling system that directly channels coolant to the motor's heat source. The coolant is injected into a hollow shaft and released over the entire area of the stator. This cooling method enables four to five times the power density compared to traditional radial-flux motors. Unlike other cooling methods that only provide a coolant jacket around the motor, Infinitum's system ensures that coolant reaches the centre of the stator. Centrifugal force then pushes the liquid to an outer jacket, where it drains into a sump for recirculation.

The resulting motor boasts impressive performance metrics, with a continuous power output of 150 kW (201 hp) and a peak power output of 300 kW (402 hp). It operates with 95% efficiency at 7500 RPM. Additionally, the motor is capable of receiving over-the-air updates, allowing for potential optimisations and enhancements in the future.

Figure 6.19 An exploded look at Infinitum's motor shows its exceptionally thin PCB stators (Source: Infinitum)

DeepDrive

DeepDrive is a startup focused on developing a dual-rotor electric motor to enhance the range of electric vehicles (EVs) in a cost- and resource-efficient manner. Their radial flux dual-rotor motor incorporates power electronics and is designed to be highly flexible, allowing for installation in various vehicle configurations, such as central drive or in-wheel drive systems.

One of the key aspects of DeepDrive's technology is its patented design, which not only extends the range of EVs but also offers a high-torque density. The motor's manufacturing process is cost-effective and utilises fewer natural resources, leading to a reduced environmental impact compared to traditional motors.

DeepDrive's electric motors boast high efficiency and provide significant benefits in terms of weight, cost and space. They are poised to enable the development of the next generation of efficient and resource-saving electric vehicles. Furthermore, the design of DeepDrive's e-motor technology aims to facilitate easy and cost-effective mass production.

Figure 6.20 BMW DeepDrive (Source: BMW Media)

Audi liquid cooled motors

In the Audi e-tron electric vehicles, the asynchronous motors utilise lightweight rotors with in-casted aluminium squirrel cages.

Figure 6.21 Operating principle of the Audi e-tron asynchronous motors: www.audi-mediacenter.com/en/audimediatv/video/audi-e-tron-electromagnetics-e-engine-animation-4843

147

These motors operate based on the principle of induction, where the current in the copper windings of the stator generates a rotating magnetic field that interacts with the rotor.

The rotor, which consists of the aluminium squirrel cage, always rotates slightly slower than the rotating field of the stator. This speed difference causes a current to be induced in the squirrel cage. As a result, a magnetic field is created in the rotor, which closes via the stator. This magnetic field exerts a circumferential force on the rotor, ultimately causing it to rotate. The animation demonstrates how the pulsating three-phase alternating current (AC) from the power electronics generates the rotating magnetic field necessary for motor operation.

In the Audi e-tron 55, this motor configuration is employed in the front e-axle, providing a maximum torque of up to 309 Nm. In the sporty Audi e-tron S, the twin rear e-axle features two of these motors, combining to deliver a maximum torque of up to 618 Nm.

The power electronics (PE) is one of the most important components of an e-axle. It converts the DC current from the battery into AC current for the motor. The pulsating three-phase current with variable frequency and amplitude generates a rotating magnetic field in the stator of the motor, which causes the rotor to turn. In case of the Audi e-tron, all functions needed to drive the motor are packed into this small box:

▶ size: 5,5 l
▶ weight: 8 kg
▶ voltage level 150 V – 460 V

Figure 6.22 Technology of electric drives: www.audi-mediacenter.com/en/audimediatv/video/audi-e-tron-power-electronics-e-engine-animation-4845

Figure 6.23 Cooling and heat dissipation of the twin-coax drive: www.audi-mediacenter.com/en/audimediatv/video/e-tron-twin-motor-animation-water-cooling-5066

▶ max. current: 530 A
▶ power density: 30 kW/l

The power density of power electronics and electric motors will continue to increase.

The animation shows the Audi e-tron 55 front e-axle with power electronics, motor and gearbox and focuses on the effective cooling concept with rotor internal cooling. The rear e-axle with coaxial architecture has a water-cooled rotor shaft.

6.3 Control system

6.3.1 Introduction

Figure 6.24 shows a generic block diagram of a PHEV. Remove the AC mains block, and it becomes an HEV, or remove the internal combustion engine (ICE), and it becomes a pure EV. The control components are outlined in the next sections. These are microprocessor control units that are programmed to react to inputs from sensors and from the driver.

6.3.2 Basic operation

To operate the motor, different phases are switched on at different times to cause rotation. The signals used to switch the transistors (usually IGBTs) are pulse-width modulated for finer control. The basic sequence is shown as Figure 6.25.

This process is examined in more detail later, but first we need to examine the stages before motor control takes place.

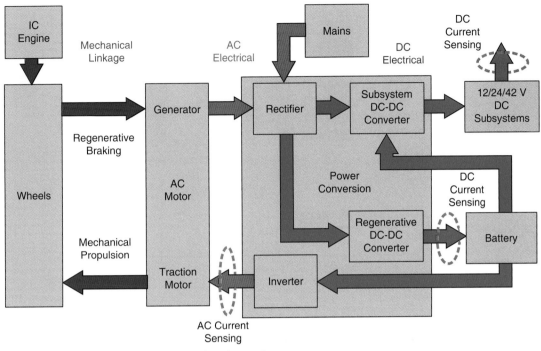

Figure 6.24 EV block diagram showing the main components

Table 6.1 Main components

Component	Purpose
Motor/generator	Provides drive to the wheels and generates electricity when the vehicle is slowing and braking
Inverter	A device to convert DC into AC
Rectifier	A device to convert AC into DC (the inverter and the rectifier are usually the same component)
DC–DC converter: regenerative	This converts the AC from the motor during braking after it has been rectified to DC. The conversion is necessary to ensure the correct voltage level for charging.
DC–DC converter: subsystem	A device to convert high-voltage DC into low-voltage DC to run the general vehicle electrics
DC subsystems	The 12 V (or 24/42 V) systems of the vehicle such as lights and wipers – this may include a small 12 V battery
Battery (high-voltage)	Usually lithium-ion or nickel–metal hydride cells that form the energy store to operate the drive motor
Battery control	A system to monitor and control battery charge and discharge to protect the battery as well as increasing efficiency
Motor control	Arguably the most important controller, this device responds to sensor signals and driver input to control the motor/generator during the various phases of operation (e.g., accelerating, cruising, braking)
Internal combustion engine	Internal combustion engine used on HEVs and PHEVs only – it hybrids with the motor. On a REV, the engine would drive a generator to charge the high-voltage battery only.

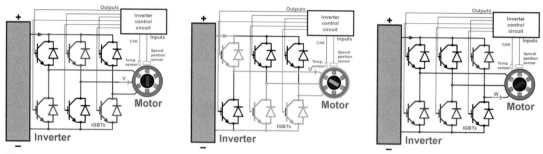

Figure 6.25 Inverter switching operation and motor rotation

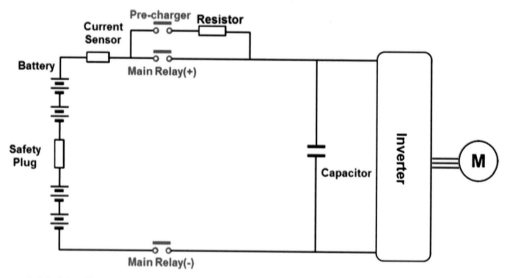

Figure 6.26 Simplified EV high-voltage circuit

6.3.3 **Start up and shut down**

Electric and hybrid vehicles are switched on by turning the key or pressing a button, just like ICE vehicles. However, the process that goes on behind the scenes is quite different. The description here is typical of most systems, but manufacturers do things in different ways so always check their information before assuming it is safe. Under normal operating conditions, the high-voltage system can be in four different conditions or modes:

▶ off
▶ start-up
▶ ready
▶ shutdown

Figure 6.26 shows a simplified high-voltage system circuit that uses three relays (or contactors):

▶ main +
▶ main –
▶ pre-charge

Off: From the off position, the relays are operated in a defined sequence. Figure 6.27 shows this sequence and the system voltages.

Start-up: When the vehicle is first switched on, the pre-charge contacts close for a short time and then open again. This allows tests to be carried out as no current should be flowing. Next the main negative contacts are closed; again, this allows circuit testing.

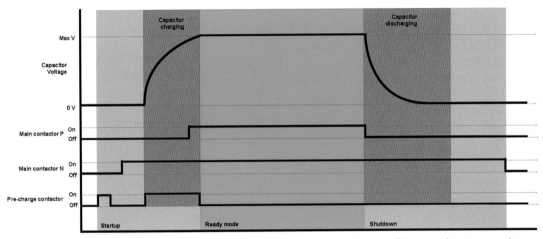

Figure 6.27 Start-up, ready mode and shutdown sequence contact positions and system voltages

The pre-charge contacts close again, and the capacitor starts to charge from the high-voltage battery through a resistor to limit the current. Finally, the main positive contacts are closed, the capacitor fully charges and the system is now described as being in *Ready* mode.

The pre-charge feature is to protect the battery as well as allowing diagnostic functions. If the capacitor was connected immediately by the main contacts, it would charge very quickly and the sudden, albeit very short, battery discharge current would be massive and could damage the cells.

> ### Key Fact
> The pre-charge feature is to protect the battery as well as allowing diagnostic functions.

The main capacitor is responsible for keeping the voltage constant and for smoothing out any voltage peaks. It can have a rating in the region of 500 mF – which is a very high value, meaning it can store a lot of energy. Touching the capacitor is potentially more dangerous than the battery, but both can kill you quickly!

Ready: The two main contacts remain closed and the vehicle can be used as normal.

'Ready', or similar, is normally displayed on the instrument panel.

Shutdown: When the vehicle is switched off, the main positive contact is opened, and the capacitor starts to discharge. Once fully discharged, the system is considered to be off. The discharge happens in two different ways:

▶ active
▶ passive

Active discharging functions by reducing the high-voltage to less than 60 V within four seconds. To do this, the transistors in the inverter are pulsed, which effectively discharges the capacitor through the motor windings. Active discharging is carried out in the event of

▶ ignition off
▶ in the event of a crash (belt tensioner or airbag triggered)
▶ pilot line open

In the event of a defective inverter, an emergency capacitor discharge is carried out using a discharge resistor, which takes about four seconds. Passive discharging is implemented with the aid of various resistors within the power and control electronics between high-voltage positive and high-voltage negative. This takes about 120 seconds, and it is carried out at all times.

On most systems, the transistors in the inverter are opened when the vehicle is being towed. If the ignition is off, the vehicle can be pushed at walking pace. If the ignition is on and the selector lever is set to the N position, then most vehicles can be towed at up to approximately 50 km/h.

If the vehicle is pushed at a speed greater than walking pace when the ignition is off or is towed at a speed greater than approximately 50 km/h with the ignition on, the transistors switch to active short circuit. The three phase lines U, V and W are short-circuited by closing the transistors. The electric drive motor can then only be rotated with a very high mechanical resistance. If the vehicle is towed for an extended period with an active short circuit, there is a risk of overheating.

6.3.4 High-voltage DC relays

High-voltage DC (HVDC) relays are used to supply and cut off the DC power, by opening and closing contacts using an actuator. Arcing that occurs when the DC power is cut off may cause damage to the contacts and surrounding components, so the arc should be extinguished as quickly as possible in the desired direction. As an example, from the company LSIS, its HVDC relay range has excellent electrical durability, compact size, with low noise and features a permanent magnet and hydrogen inside for optimal arc extinguishing. Hydrogen is only flammable if oxygen is also present.

> **Key Fact**
> Arcing occurs when DC power is cut off, and this can cause damage to relay contacts.

The company LSIS has been developing its 450 V devices and now have 1000 V and 1500 V relays. Generally, as the rated voltage increases, the size of the product must be increased to ensure insulation. However, the new series is implemented in the same size as the earlier types, thus improving the space utilisation.

As the battery capacity of the vehicle increases, the rated voltage is increased and the external charge is expanded. Therefore, it is required to monitor the increase of the DC relay voltage and the on/off of the contact point. The GPR-M series (Figure 6.28), for example, has a rated voltage of DC 1000 V, 10 A to 400 A.

6.3.5 Power control

Motor/generator control: The motor/generator control system mainly performs motor control to provide drive as well as regeneration when the motor is acting as a

Figure 6.28 GPR-M series relays (Source: LSIS)

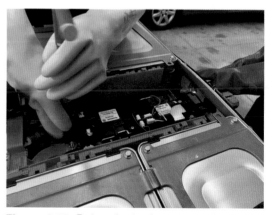

Figure 6.29 Relays in the battery pack

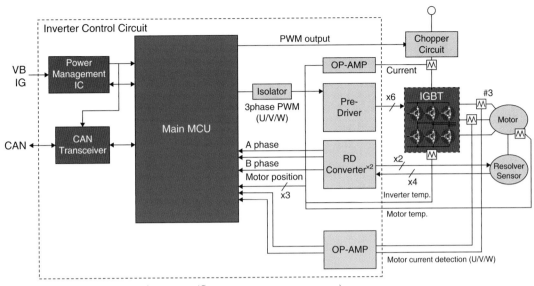

Figure 6.30 Motor control system (Source: www.renesas.eu)

generator. The main microprocessor control unit (MCU) controls the inverter via a pre-driver circuit. The sequence in which the inverter (labelled IGBT in Figure 6.30) is switched and at what rate determines the torque and speed of the motor.

> **Key Fact**
>
> The main microprocessor control unit (MCU) controls the inverter.

The insulated-gate bipolar transistor (IGBT) is a three-terminal power semiconductor device primarily used as a fast-acting, high-efficiency electronic switch. It is used to switch electric power in many modern appliances as well as electric vehicles.

Inverter: The electronic circuit used to drive a motor is usually called an inverter because it effectively converts DC to AC. An important aspect of this type of motor and its associated control is that it works just as effectively as a generator for regenerative braking. It is controlled by the main MCU in the motor controller. The switches shown in Figure 6.30 will in reality be IGBTs. The IGBTs, in turn, are

Figure 6.31 IGBTs

controlled by a pre-driver circuit that produces a signal that will switch the inverter in a suitable sequence.

> **Key Fact**
>
> An inverter converts DC to AC.

The switching actions are shown in Figure 6.32 and the output signal from the inverter when

153

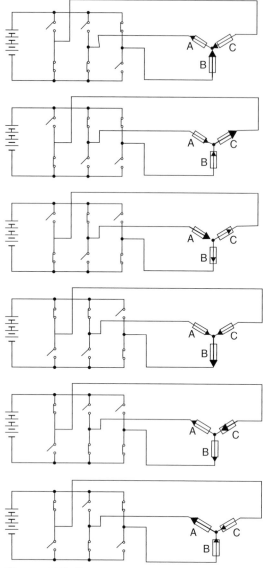

Figure 6.32 Representation of the inverter switching operation

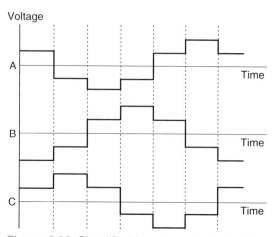

Figure 6.33 Simplified inverter output signals

> **Definition**
>
> Pulse-width modulation (PWM):
> A method of reducing the average power delivered by an electrical signal, by chopping it up into parts. The average value of voltage fed to the load is controlled by turning the switch on and off at a fast rate.

it is driving the motor is shown, in simplified format, in Figure 6.33.

6.3.6 Pulse-width modulation

Pulse-width modulation (PWM) is the process used to control the switching components (IGBTs or MOSFETs) and, therefore, to drive the motor. The switching efficiency using this method usually exceeds 90%.

PWM controls the power switch output power by varying its ON and OFF times. The ratio of ON time to the switching period time is the duty cycle. Figure 6.34 shows how the duty cycle is varied (blue trace) to produce a sinewave or effectively and AC output. The higher the duty cycle, the higher the power semiconductor switch output power.

Figure 6.34 shows just one output. In an EV inverter, this is carried out three times, 120 degrees apart, so as to produce a three-phase output as shown in Figure 6.35. In this image, the output voltage is being varied. This is an actual waveform captured using a PicoScope with an amp clamp on phases U, V and W, which are connected from the inverter to the motor.

6.3.7 Sensors

In order to carry out motor/generator switching accurately, the power/control electronics need to

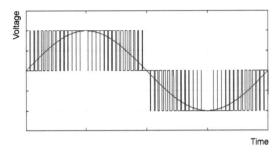

Figure 6.34 Inverter switching pattern used to generate three-phase AC from a DC supply

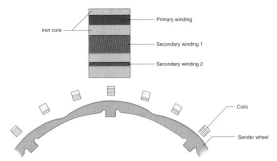

Figure 6.36 An inductive resolver provides speed and position information

know the condition and exact speed and position of the motor. The speed and position information is supplied by one or more sensors mounted on the casing in conjunction with a reluctor ring.

Key Fact

Power/control electronics need to know the condition and exact speed and position of the motor.

The system shown in Figure 6.36 uses 30 sensor coils and an eight-lobe reluctor ring. The output signal changes as a lobe

approaches the coils, and this is recognised by the control unit. The coils are connected in series and consist of a primary and two secondary windings around an iron core. The separate windings produce different signals (Figure 6.37) because as the reluctor moves, it causes the signal in each secondary winding to be amplified. The position of the rotor can, therefore, be determined with high accuracy using the amplitudes of the signals. The frequency of the signal gives the rotational speed.

A drive motor temperature sensor is usually used, and it also sends signals to the electric

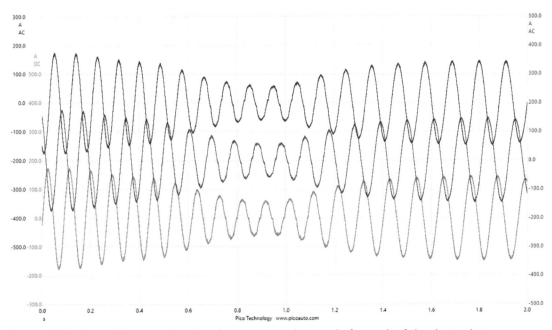

Figure 6.35 Current/time graphs showing one complete cycle for each of the three phases

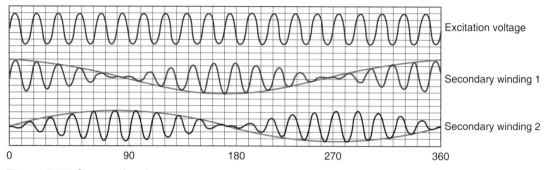

Figure 6.37 Sensor signal

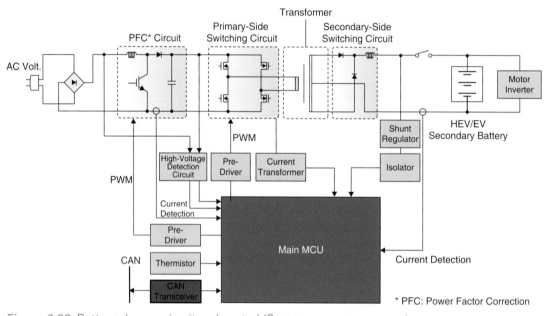

Figure 6.38 Battery charger circuit and control (Source: www.renesas.eu)

drive control unit. Typically, the power of the drive motor is restricted at temperatures above about 150 °C, and, in some cases above 180 °C, it may even prevent the drive from being used to protect it against overheating. The sender is normally a negative temperature coefficient (NTC) thermistor.

6.3.8 Battery

Battery charger: A charger and DC–DC step-up system controls the AC input from a household power supply and boosts the voltage, using a DC–DC converter, to whatever is required by the battery. The MCU performs power factor correction and control of the DC–DC step-up circuit (Figure 6.38).

> **Key Fact**
> Power factor correction (PFC) is an energy-saving technology that is used to improve the operating efficiency of electrical power systems.

Battery control: A battery control system is used for managing remaining battery voltage and control of battery charging. Voltages of individual cells are monitored and the balance

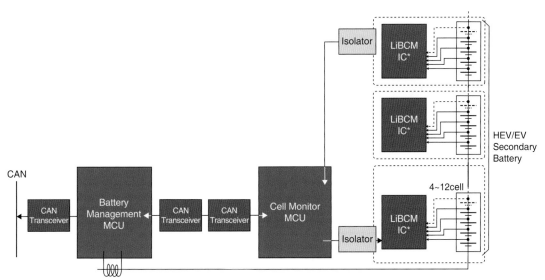

* LiBCM IC: Li-ion Battery Cell Management IC

Figure 6.39 Battery control system (Source: www.renesas.eu)

controlled by a battery cell monitor MCU and lithium-ion battery cell monitor integrated circuits (Figure 6.39).

6.3.9 Battery management systems

In electric vehicles (EVs), the battery pack requires continuous monitoring to ensure optimal performance and prevent any potential failures. This monitoring is achieved through a battery management system (BMS), which is an electronic system responsible for managing the charging and discharging of the battery cells.

The primary function of the BMS is to protect the battery by monitoring various parameters such as temperature and voltage signals from the individual cell modules as well as pack-level current signals. It constantly checks these signals to ensure that the battery operates within its safe operating limits. If any parameter exceeds the predefined thresholds, the BMS takes appropriate actions to mitigate potential risks.

The BMS also facilitates the balance or control of the module environment. This involves managing the state of charge of individual cells within the battery pack as well as ensuring that each cell operates uniformly. By balancing the cells, the BMS helps optimise the overall performance and longevity of the battery pack.

To achieve its functions, the BMS requires accurate and reliable operation. It needs to be compact and lightweight, minimising the additional bulk it adds to the battery pack. The BMS is typically connected to a battery management controller (BMC), which receives and processes the monitored signals. The processed signals are then transmitted to the cell management controllers (CMC), which are responsible for balancing the cells and enabling controlled power flow, especially during charging.

Connectivity is crucial for transferring signals between the various components of the BMS. This allows for real-time monitoring, analysis and control of the battery pack, ensuring its safe and efficient operation.

6.3.10 Cell balancing

High-quality lithium-ion cells have a very uniform capacity and low self-discharge rate when they are new. However, balancing

157

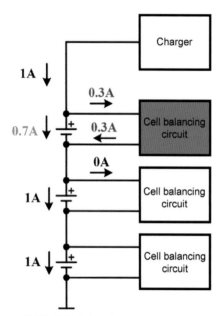

Figure 6.40 Principle of cell balancing

is advantageous because the performance of individual cells can decrease at a slightly different rate. If just one cell in a series connection loses capacity or develops increased self-discharge, the whole string is affected.

Passive balancing uses a resistor that can be connected across a cell to divert some of the charge current. This is usually done when the cells are in the range of 70–80% charged. Active balancing takes any extra charge from higher voltage cells during discharge to those with a lower voltage.

> ### Key Fact
> Passive balancing uses a resistor that can be connected across a cell to divert some of the charge current.

Active balancing is the method used by manufacturers for many EV batteries. It is more complex than passive balancing as it requires DC–DC converters but is more beneficial overall. The corrections to charging current are in the mA range. Because of

the heavy loads during acceleration, which are often followed by fast charging from regenerative braking, it is essential to keep the cells balanced to extend battery life.

Passive balancing is a method used to equalise the state of charge (SOC) among the cells in a battery pack. It typically operates at fixed points, either "top balanced", where all cells reach 100% SOC simultaneously, or bottom balanced, where all cells reach the minimum SOC at the same time.

In passive balancing, the goal is to transfer energy from cells with a higher state of charge to cells with a lower state of charge. This is accomplished by either bleeding energy from cells with higher SOC, such as through a controlled short circuit using a resistor or transistor, or by shunting energy through a parallel path during the charging cycle. By diverting some of the current away from cells that are already charged, passive balancing ensures that less of the charging current is consumed by those cells.

However, passive balancing is inherently inefficient as it results in energy loss in the form of heat. The wasted energy is dissipated as heat in the balancing process, which can be a limitation for rapid balancing. The buildup of waste heat can potentially affect the rate at which balancing can occur.

It's important to note that passive balancing can only take place during steady charging of the battery. It relies on the charging process to redirect current flow and equalise the SOC of the cells.

Overall, while passive balancing helps in achieving a more balanced SOC among battery cells, it has limitations in terms of energy efficiency and the speed at which it can balance the cells. Active balancing methods, such as using active electronic circuits, offer more efficient and precise balancing, but they typically require additional components and control algorithms.

Active balancing is an alternative method used to redistribute excess energy from cells with

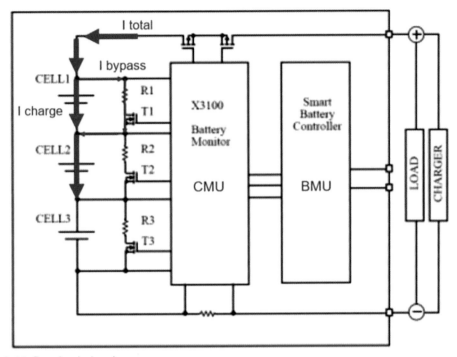

Figure 6.41 Passive balancing

larger capacity to cells with lower capacity within a battery pack. Unlike passive balancing, active balancing actively transfers energy between cells to achieve a more balanced state of charge (SOC) and make better use of the cells' capacity.

In active balancing, excess energy from cells with a higher SOC can be transferred to cells with a lower SOC through different techniques. One method involves switching a reservoir capacitor in-circuit with the cell, allowing it to absorb excess energy. The capacitor is then disconnected and reconnected to a cell with lower SOC, transferring the stored energy to balance the cells. Another approach is to use a DC-to-DC converter connected across the entire battery pack, which can redistribute energy between cells with different SOC levels.

Although some energy is still wasted as heat in active balancing, it is typically less wasteful compared to passive balancing. Active

balancing offers more efficient utilisation of the cells' capacity by taking advantage of the cells with higher capacity.

It's worth noting that active balancing can occur during both charging and discharging of the battery. This flexibility allows for continuous balancing and maintenance of cell SOC levels.

However, active balancing does come with additional costs and complexities. The implementation of active balancing requires additional components such as capacitors, switches or DC-to-DC converters. It also requires control algorithms and circuitry to manage the balancing process effectively. These added complexities and costs may not always be justified depending on the specific application or the characteristics of the battery pack.

In summary, active balancing provides more efficient redistribution of energy between cells, leveraging the capacity of more capacious

159

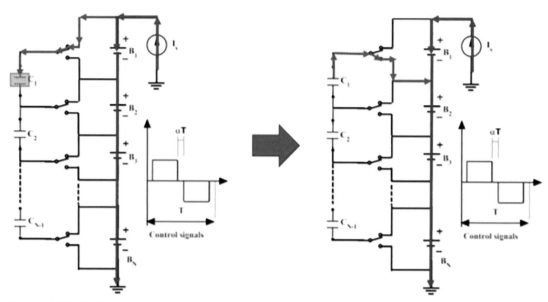

Figure 6.42 Active balancing

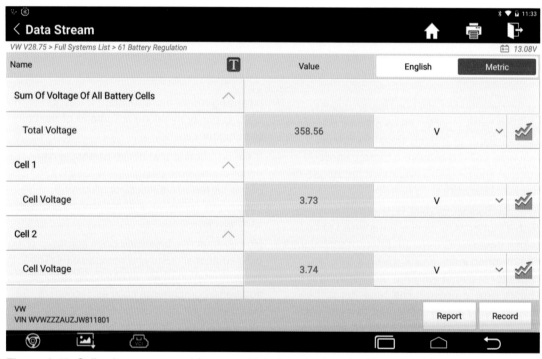

Figure 6.43 Cell voltages scanned from a vehicle (good values in this example)

cells. It can occur during both charging and discharging. However, the decision to use active balancing should consider the cost, complexity and overall benefits specific to the application at hand.

6.3.11 Component cooling

To protect the sensitive components from high temperatures, the correct temperature is often maintained using coolant. The coolant

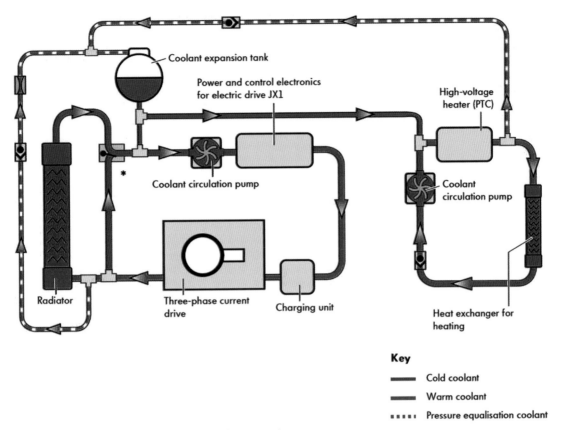

Coolant expansion tank

Power and control electronics
for electric drive JX1

High-voltage
heater (PTC)

Coolant circulation pump

Coolant
circulation pump

*

Radiator

Three-phase current
drive

Charging unit

Heat exchanger for
heating

Key

—— Cold coolant

—— Warm coolant

▪ ▪ ▪ ▪ ▪ Pressure equalisation coolant

Figure 6.44 Cooling circuit (Source: Volkswagen)

temperature can be as high as 65 °C and is electronically monitored and regulated by the motor control unit. The following components are cooled using coolant:

▶ three-phase current drive

▶ charging unit
▶ power and control electronics

Figure 6.44 shows a cooling circuit from a Volkswagen car. It is typical of many.

CHAPTER 7

Charging

7.1 Charging, standards and infrastructure

7.1.1 Infrastructure

As of 1 July 2023, there were 44,020 public electric vehicle charging devices installed in the UK, within which

- ▶ 8,461 were rated 'rapid' devices or above, this represents 19% of all charging devices
- ▶ 24,918 were rated 'fast' chargers, this represents 56% of all charging devices
- ▶ 21,294 were designated as 'destination' chargers, this represents 47% of all charging devices
- ▶ 14,848 were designated as 'on street' chargers, this represents 34% of all charging devices[1]

Rapid charging or above devices are those whose fastest connector is rated at 25 kW and above and includes charge points classified as either rapid or ultra-rapid

- ▶ slow-charging devices represent 3 kW to 6 kW
- ▶ fast-charging devices represent 7 kW to 22 kW

- ▶ rapid-charging devices represent 25 kW to 100 kW
- ▶ ultra-rapid-charging devices represent 100 kW+

Most electric cars will be charged at home, but as noted, national infrastructures are developing. There are, however, competing organisations and commercial companies, so it is necessary to register with a few different organisations to access their charging points. Some are pay-in-advance, some are pay-as-you-go and others require a monthly subscription. Many apps and websites are available for locating charge points. One of the best I have found is www.zap-map.com. Many businesses now also provide charging stations for staff and visitors.

Although rechargeable electric vehicles and equipment can be recharged from a domestic wall socket, a charging station has additional current or connection-sensing mechanisms to disconnect the power when the EV is not charging. There are two main types of safety sensor:

1. Current sensors monitor the power consumed and only maintain the connection if the demand is within a predetermined range.

DOI: 10.1201/9781003431732-7

2. Additional sensor wires provide a feedback signal that requires special power plug fittings.

The majority of public charge points are lockable, meaning passers-by cannot unplug the cable. Some charge points can send a text message to the car owner if the vehicle is unexpectedly unplugged or tell you when the vehicle is fully charged.

Safety First

Current sensors monitor the power consumed and maintain the charging connection if the demand is within a predetermined range.

It is safe to charge in wet weather. When you plug in the charge lead, the connection to the supply is not made until the plug is completely in position. Circuit breaker devices are also used for additional safety. Clearly some common sense is necessary, but EV charging is very safe.

Key Fact

When you plug in the charge lead, the connection to the supply is not made until the plug is completely in position.

Domestic charge points: It is strongly recommended that home charging sockets and wiring are installed and approved by a qualified electrician. A home charge point with its own dedicated circuit is the best way of charging an EV safely. This will ensure the circuit can manage the electricity demand from the vehicle and that the circuit is activated only when the charger communicates with the vehicle, known as the 'handshake'. For rapid charging, special equipment and an upgraded electrical supply would be required and is, therefore, unlikely to be installed at home, where most consumers will charge overnight.

Key Fact

Most consumers will charge overnight, but charging from solar panels will also become popular.

Figure 7.2 Charging point (Source: Richard Webb, www.geograph.ie)

Figure 7.1 Charging point on the roadside (Source: Rod Allday)

Figure 7.3 Charging at home

Figure 7.4 Domestic charging point

7.1.2 **Standardisation**

So that electric vehicles can be charged everywhere with no connection problems, it was necessary to standardise charging cables, sockets and methods. The IEC publishes the standards that are valid worldwide, in which the technical requirements have been defined. Table 7.1 lists some of the most important standards associated with the charging of EVs.

7.1.3 **Charging time and cost**

How long it takes to charge an EV depends on the type of vehicle, how discharged the battery is and the type of charge point used. Typically, pure electric cars using standard charging will

take about 10 hours to charge fully and can be 'opportunity charged' whenever possible to keep the battery topped up. But it is important to note that this varies a lot depending on the vehicle and the charger.

> **Key Fact**
>
> Charge time for an EV depends on the type of vehicle, temperature, size of the battery, how discharged the battery is and the type of charge point used.

Pure EVs capable of using rapid charge points could be fully charged in around 30 minutes and can be topped up in around 20 minutes, depending on the type of charge point and available power.

The cost of charging an EV depends on the size of the battery and how much charge is left in the battery before charging. If you charge overnight, you may be able to take advantage of cheaper electricity rates when there is surplus energy. The cost of charging from public points will vary; many will offer free electricity in the short term. It is also possible to register with supply companies who concentrate on energy from renewable sources.

Here is a simple way to calculate cost of recharging:

▶ electricity is paid for in units (1 unit is 1 kWh)
▶ charging is about 85% efficient (but often better)

To charge a large 100 kWh battery from 20% to 80% requires 60 kWh

Divide this by the efficiency (85%) and the answer is just over 70 kWh – multiply this by the cost per unit of electricity (let's say 30 p, but this will vary a lot) and it would cost about £21.

I currently buy electricity (overnight rate) at 7.5 p per unit so a recharge like this costs just over £5. Compare this with an ICE top up . . .

Table 7.1 Charging standards

IEC 62196-1	IEC 62196-2	IEC 62196-3	IEC 61851-1	IEC 61851-21-1	IEC 61851-21-2	HD 60364-7-722
Plugs, socket outlets, vehicle connectors and vehicle inlets. Conductive charging of electric vehicles	Dimensional compatibility and interchangeability requirements for AC pin and contact tube accessories. The permissible plug and socket types are described	Dimensional compatibility and interchangeability requirements for dedicated DC and combined AC/DC pin and contact tube vehicle couplers	Electric vehicle conductive charging system. Different variants of the connection configuration, as well as the basic communication with the vehicle, are defined in this standard	Electric vehicle conductive charging systems. Electric vehicle on-board charger EMC requirements for conductive connection to an AC/DC supply	Electric vehicle conductive charging systems EMC requirements for off-board electric vehicle charging systems	Low-voltage electrical installations. Requirements for special installations supply of electric vehicles

Figure 7.5 Charging forecourt (Source: Gridserve)

7.1.4 **Charging methods, modes and plugs**

There are a number of different methods and modes of EV charging explained in this section, but, at a simple level, there are two distinct types of charging:

▶ alternating current (AC)
▶ direct current (DC)

Power is supplied from the grid as AC, but an EV battery is always charged using DC. The current, therefore, needs to be converted from AC to DC, and where this takes place depends on the type of charge point. This is summarised in Figure 7.6.

AC charging: Alternating current charging has now established itself as a standard charging method. It is possible in the private sector, as well as at charging stations in the semi-public and public sector, with relatively low investments. Consequently, this charging method also has a long-term future. Standard charging occurs via an alternating current connection and is the most common and most flexible charging method. In charging Modes 1 and 2, charging is possible on household sockets or on CEE sockets. On the household socket, charging can take up to several hours due to the power limited through the socket, depending on rechargeable battery capacity, fill level and charging current.

> **Key Fact**
> Alternating current charging has now established itself as the standard charging method.

In charging Mode 3, a vehicle can be charged at a charging station where power of up to 43.5 kW is possible with a significantly reduced charging time. Particularly in the private sector, the usable power is limited by the fuse protection of the building connection. Charging powers to maximum 22 kW at 400 V AC are

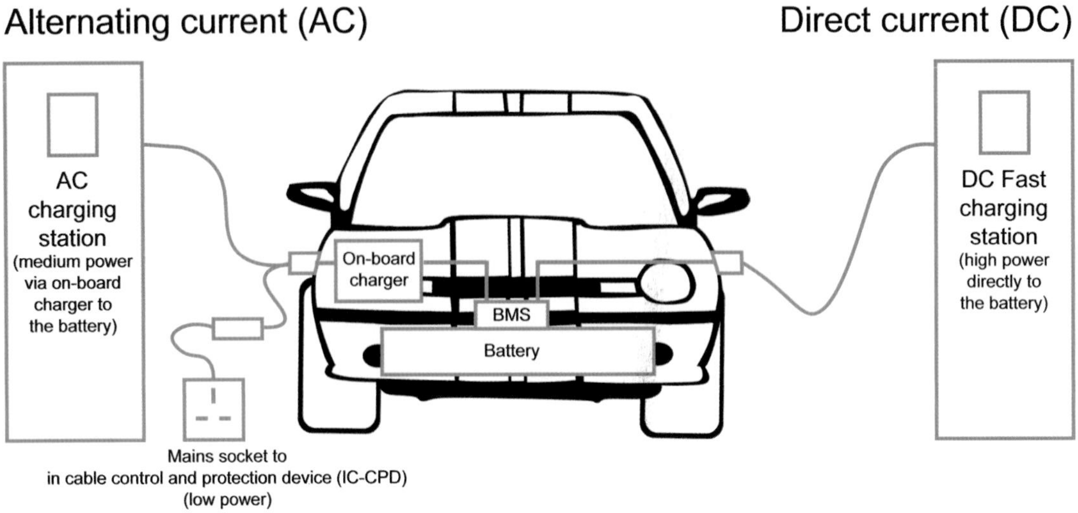

Alternating current (AC)   Direct current (DC)

Figure 7.6 AC and DC charging modes

usually the high power limit for home charging stations.

The charging device is permanently installed in the vehicle. Its capacity is adjusted to the vehicle battery. Compared with other charging methods, the investment costs for AC charging are moderate.

DC charging: With direct-current charging, there is a distinction between

▶ DC low charging: up to 38 kW with type 2 plugs
▶ DC high charging: up to 170 kW

The charging device is part of the charging station, so DC-charging stations are significantly more expensive as compared with AC charging stations. The prerequisite for DC charging is an appropriate network of charging stations, which due to the high power require high infrastructure investments. Fast charging with high currents requires appropriately dimensioned line cross-sections that make connecting the vehicles to the charging station more cumbersome. Standardisation of the DC-charging connection has not yet been concluded, and market availability is still uncertain. In practice, vehicles with a

DC-charging connection have an additional connection for standard charging so that the vehicle can also be charged at home.

Key Fact

DC-charging stations are significantly more expensive compared with AC-charging stations.

Inductive charging: Charging occurs without contacts via inductive loops. The technical complexity and, thus, the costs are considerable for the charging station as well as the vehicle. This system is not yet ready for the market or for large-scale production.

Definition

Inductive charging: no physical connection is made; instead, transformer action or mutual induction is used.

Battery replacement: The vehicle's rechargeable battery is replaced with a fully charged battery at the change station. In this

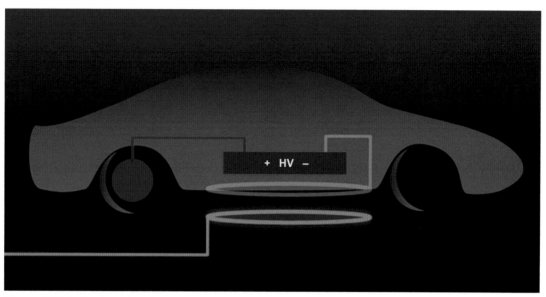

Figure 7.7 Inductive charging

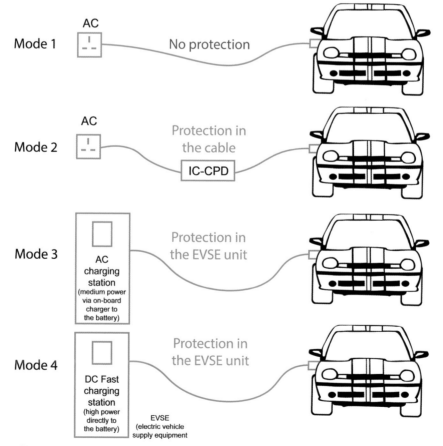

Figure 7.8 Charging modes and a summary of each

case, you can continue driving after a few minutes. The prerequisite for this concept would be that vehicle manufacturers would have to install standardised rechargeable batteries at standardised positions in the vehicle. However, such standardisation would hardly be possible due to the different vehicle types and uses. The charge stations would have to keep battery types for the different vehicles on hand, which in practice would be equally difficult. Consequently, battery replacement could only be implemented today in closed fleets.

Four different charging modes have been defined for safe charging of electric vehicles in line with demand. These charging modes differ relative to the power source used (protective contact, CEE, AC- or DC-charging socket), and they differ relative to the maximum charging power and the communication possibilities.

Safety First

Four different charging modes have been defined for safe charging of electric vehicles.

Mode 1: This mode charges from a socket to max 16 A, three-phase without communication with the vehicle. The charging device is integrated in the vehicle. Connection to the energy network occurs via an off-the-shelf, standardised plug and socket that must be fused via a residual current protective device. This method is not recommended because Mode 2 offers greater safety thanks to communication with the vehicle.

Mode 2: This mode charges from a socket to max 32 A, three-phase with a control function and protective function integrated in the cable or the wall-side plug. The charging device is installed in the vehicle. Connection to the energy network occurs via an off-the-shelf, standardised plug and socket. For Mode 2, the standard prescribes a mobile device to increase the level of protection.

Moreover, for the power setting and to satisfy the safety requirements, a communication device is required with the vehicle. These two components are combined in the in-cable control box (ICCB).

Mode 3: This mode is for charging at AC-charging stations. The charging device is a fixed component of the charging station and includes protection. In the charging station, PWM communication, residual current device (RCD), overcurrent protection, shutdown and a specific charging socket are prescribed. In Mode 3, the vehicle can be charged three-phase with up to 63 A, so a charging power of up to 43.5 kW is possible. Depending on rechargeable battery capacity and charge status, charges in less than one hour are possible.

Mode 4: This mode is for charging at DC-charging stations. The charging device is a component of the charging station and includes protection. In Mode 4, the vehicle can be charged with two plug-and-socket systems, both of which are based on the type 2 plug geometry. The 'combined charging system' has two additional DC contacts to 200 A and up to 170 kW charging power. The other option is a plug and socket with lower capacity for a charge to 80 A and up to 38 kW in type 2 design.

Standards continue to be reviewed and changed to improve safety and ease of use as well as compatibility.

Figure 7.9 In-cable control box (ICCB)

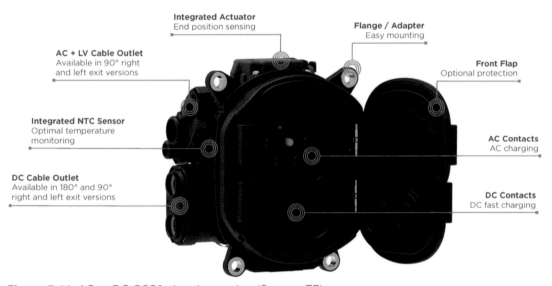

	Japan	N. America	Europe	China	All except EU
AC 3.7 - 43kW on-board charger converts AC to DC	Type 1	Type 1	Type 2	GB/T	
DC 50 - 150kW fast charging (no active cooling) 200 - 500kW high power (requires active cooling charging station	CHAdeMO	CCS 1	CCS 2	GB/T	Tesla

Figure 7.10 Current and future charging methods

Integrated Actuator
End position sensing

Flange / Adapter
Easy mounting

AC + LV Cable Outlet
Available in 90° right
and left exit versions

Front Flap
Optional protection

Integrated NTC Sensor
Optimal temperature
monitoring

AC Contacts
AC charging

DC Cable Outlet
Available in 180° and 90°
right and left exit versions

DC Contacts
DC fast charging

Figure 7.11 AC or DC CCS2 charging socket (Source: TE)

Figure 7.10 compares different charging methods (standard and proprietary).

Safety First

Voltage is only switched on when the system has detected that the plugs on the vehicle side and infrastructure side are completely plugged in, that the plugs are locked and that the protective conductor connection is correct.

7.1.5 Communication

Basic communication: Safety check and charging current limitation is determined. Even before the charging process starts, in charging Modes 2, 3 and 4, PWM communication with the vehicle occurs via a connection known as the control pilot (CP) line. Several parameters are communicated and coordinated. The charging will only begin if all security queries clearly correspond to the specifications and the maximum permissible charging current

Table 7.2 Charging plugs

Type of plug	Image
Type 1 is a single-phase charging plug developed in Japan exclusively for the vehicle side-charging connection. The maximum charging power is 7.4 kW at 230 V AC. Type 1 offers insufficient possibilities for the three-phase European networks. This plug is defined by SAE J1772 standard and is also known as a J plug. It is a North American standard.	 **Figure 7.12** Type 1 plug (J plug in the USA)
Type 2 The IEC 62196 Type 2 connector (often referred to as Mennekes) is used mainly within Europe, as it was declared standard by the EU. The connector is circular in shape, with a flattened top edge; the original design specification carried an output electric power of 3–50 kW using single-phase (230 V) or three-phase (400 V) alternating current (AC), with a typical maximum of 32 A 7.2 kW using single-phase AC and 22 kW with three-phase AC. The plugs have openings on the sides that allow both the car and the charger to lock the plug automatically to prevent unwanted interruption of charging or theft of the cable.	 **Figure 7.13** Type 2 plug
CHAdeMO stands for CHArge de MOde and was developed in 2010. The latest CHAdeMO Protocol (3.0) allows for up to 900 kW of charging (600 A x 1.5 kV) through the CHAdeMO/CEC collaboration. This plug has mostly been replaced by CCS in Europe.	 **Figure 7.14** CHAdeMO plug (Source: D-Kuru)
Combined charging system (CCS) covers charging electric vehicles using the combo 1 and combo 2 connectors at up to 80 or 350 kW. These two connectors are extensions of the type 1 and type 2 connectors, with two additional direct current (DC) contacts to allow high-power DC fast charging. CCS allows AC charging using the type 1 and type 2 connector depending on the geographical region. Since 2014, the EU has required the provision of type 2 or combo 2 within the European electric vehicle network.	 **Figure 7.15** CCS combo plug type 1
Tesla's current plug in Europe is based on the type 2 Mennekes connector, which is basically identical to the top half of the CCS combo 2 plug. The current Supercharger network is being updated, while it appears that the existing type 2 plugs will fit in the top half of the CCS type 2, apparently this may not charge the Tesla.	 **Figure 7.16** CCS combo plug type 2 socket (Source: Tesla)

has been communicated. These test steps are always executed:

Definition
PWM: pulse-width modulation.

1. The charging station locks the infrastructure side-charging coupler.
2. The vehicle locks the charging coupler and requests start of charging.
3. The charging station (in Mode 2, the control unit in the charging cable) checks the connection of the protective conductor to the vehicle and communicates the available charging current.
4. The vehicle sets the charger accordingly.

If all other prerequisites are met, the charging station switches the charging socket on. For the duration of the charging process, the protective conductor is monitored via the PWM connection and the vehicle has the possibility of having the voltage supply switched off by the charging station. Charging is ended, and the plugs and sockets are unlocked via a stop device (in the vehicle).

Limitation of the charging current: The vehicle's charging device determines the charging process. To prevent the vehicle charging device from overloading the capacity of the charging station or of the charging cable, the power data of the systems is identified and adjusted to match. The CP box reads the power data of the cable from the cable. Before the charging process is started, the box communicates the power data to the vehicle via PWM signal, the vehicle's charging device is adjusted accordingly and the charging process can begin, without the possibility of an overload situation occurring.

Key Fact
The vehicle's charging device determines the charging process.

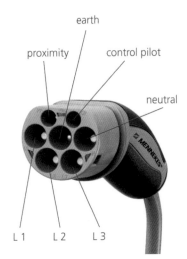

Figure 7.17 Type 2 AC charging plug and its pinout

The weakest link in the charging chain determines the maximum charging current: The charging current in the charger is limited depending on the power of the charging station and the resistance coding in the plug of the charging cable.

Key Fact
The weakest link in the charging chain determines the maximum charging current.

7.1.6 Fast charging

Before discussing fast charging further, we must repeat the general charging advice as summarised in Figure 7.18: Store cold, use hot, charge slow and store half-full!

However, fast charging from time to time is perfectly fine (but check the manufacturer's recommendations as this may affect the battery warranty). Figure 7.19 shows typical charge times as related to the power available from the charge and assumes the vehicle is able to accept the full amount.

It is okay to use a DC fast charger when you are on a trip and need to recharge your EV in a half hour. But manufacturers suggest

Store cold, use hot

Charge slowly

Store half full

Figure 7.18 Prolonging the life of a battery (in your phone and car)

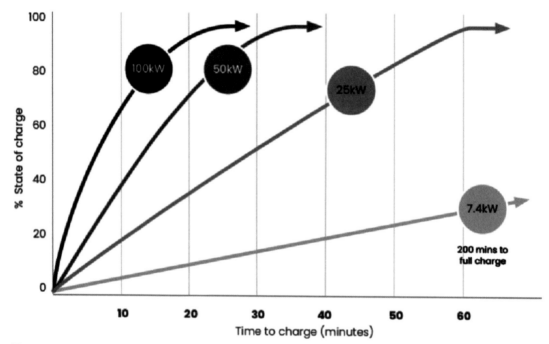

Figure 7.19 Charge times

it is best not to use Level 3 charging on a daily basis.

Battery degradation is caused by loss of active lithium and increased internal resistance. The factors involved in this are

▶ temperature
▶ depth of discharge
▶ charging rates
▶ battery cell design
▶ age

Fast charging (excessively) can cause plating as shown in Figure 7.20. Again, we would stress fast charging is fine when needed – just limit it where possible!

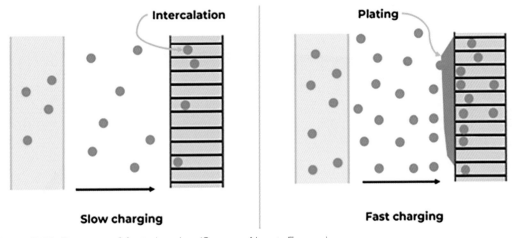

Intercalation

Plating

Slow charging

Fast charging

Figure 7.20 Dangers of fast charging (Source: About: Energy)

7.1.7 Vehicle-to-grid

Vehicle-to-grid (V2G) is a system that uses bidirectional power from the car to the grid as well as the normal charging routine of grid to car. If this system is employed, the car battery can be used as a power backup for the home or business. If the car is primarily charged from renewable sources such as PV panels or wind generation, then returning this to the grid is not only ecologically beneficial but also an ideal way of stabilising fluctuations of demand in the grid. The potential problem is managing inrush currents if lots of vehicles fast charge at the same time. This notion is still a little way into the future at the time of writing (2023), but the concept of the 'smart grid' using techniques such as this is not far off.

However, one key stumbling block to the V2G technology is the battery warranty. Most battery manufacturers guarantee their batteries for a set time (several years) but have a proviso that there is a limit on how much it is used because use is increased with V2G. This is often described as 'coulombs in and coulombs out'. The coulomb (C) is the SI unit of electric charge. It is the charge (Q) transported by a constant current of one ampere in one second.

> **Definition**
> The coulomb (C) is the SI unit of electric charge. It is the charge (Q) transported by a constant current of one ampere in one second.

7.1.8 Charging case study

The lithium-ion battery system of the Audi e-tron GT quattro and the RS e-tron GT can store 84 kWh of energy net (93 kWh gross). It integrates 33 cell modules, each of which comprises 12 pouch cells with flexible outer skin. Each module is fitted with its own computer that monitors the temperature and voltage. The unusually high system voltage of approximately 800 V enables a powerful continuous output and shortens the charging duration; in addition, it reduces the weight of and space required by the wiring.

This battery system is located beneath the passenger compartment, at the lowest point of the car. This, in combination with the electric motors, provides a low centre of gravity appropriate for a sports car and a weight distribution between the front and rear axles that is very close to the ideal value of 50:50. Thirty modules form the lower level

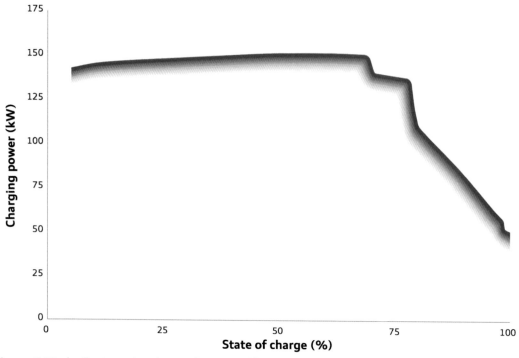

Figure 7.21 Audi e-tron charging performance (Source: Audi Media)

of the battery, which features a wide recess in the rear section. It creates space for the rear passengers' feet, which allows them to sit in a low position and also enables the flat vehicle silhouette. The upper floor contains three further modules situated below the rear seats. The connections, fuses and the main control unit are located under the console of the centre tunnel. The bottom of the battery is protected by an aluminium plate.

The inner structure of the battery that houses the module, the frame surrounding it and the upper cover plate are made of aluminium. As in the body, die-cast sections, extruded sections and aluminium sheets are used here. The battery system contributes significantly to the rigidity of the body to which it is attached via 28 screws. At the same time, it improves passive safety in the event of a front and side impact.

Beneath the cell space of the battery, there is a compound structure of flat extruded sections through which a glycol/water mixture flows that

circulates in its own low-temperature circuit. The temperature is transferred between the cooling plates and the cell space via a heat-conducting paste. The battery's feel-good temperature is between 30 °C and 35 °C, and its operating range extends from −30 °C to 50 °C.

Four separate coolant circuits, each at its own temperature level, regulate the temperature in the high-voltage components and the interior precisely and quickly. They can be interconnected flexibly as required. If the driver demands a high output several times in a row, valves couple the coolant circuit of the battery with the refrigerant circuit of the air-conditioning system – this intensive cooling keeps the performance of the drive at a consistently high level. The e-tron GT quattro and RS e-tron GT can accelerate to full speed from a standstill up to 10 consecutive times.

The refrigerant circuit also helps with cooling during fast DC charging, which can heat the battery up to 50 °C. The thermal management

Figure 7.22 Audi battery and system

is connected to the navigation system. When the driver sets a high power charging (HPC) terminal as the destination, the cooling of the battery is already intensified on the way to the charging station so that it can be charged as quickly as possible. Should the battery still be very cold shortly after the car is started in winter, it is heated for fast charging.

The standard equipment of the e-tron GT includes a heat pump that heats the interior with the waste heat of the high-voltage components. It can reduce the loss of range that the electric climate control causes in winter, in particular, significantly. In addition to charging, customers can also manage pre-entry climate control of the interior via their smartphones using the myAudi app. This is done via a powerful high-voltage heating element and does not depend on the car charging via the power grid. Audi equips the e-tron GT with a deluxe auxiliary air-conditioning system as an option that also incorporates the steering wheel rim (if heatable), the exterior mirrors and the rear window.

The charging flaps of the GT are located behind the front wheels. Both sides feature connections for alternating current (AC), and there is also a connection for direct current (DC) on the right-hand side. The Audi e-tron GT is delivered to its customers with two charging cables as standard: one Mode 3 cable for public AC terminals and the charging system compact for the garage. The e-tron GT can charge with 11 kW AC as standard, which allows it to recharge an empty battery overnight. An optional on-board charger for 22 kW will follow shortly after the market launch.

At a direct-current terminal with a voltage of 800 V, for example, in the European motorway network from Ionity, the Audi e-tron GT achieves a peak charging capacity of up to 270 kW. This allows it to recharge energy for up to 100 kilometres (62.1 mi) in just over five minutes and charging from 5% to 80% SoC (state of charge) takes less than 23 minutes under ideal conditions. The driver can restrict the charging target if the rate appears too high.

7.2 **Wireless power transfer**

7.2.1 **Introduction**

Range anxiety continues to be an issue to EV acceptance. Wireless power transfer (WPT) is a means to increase the range of an electric vehicle without substantial impact on the weight or cost. WPT is an innovative system for wirelessly charging the batteries in electric vehicles. There are three categories:

▶ stationary WPT: Vehicle is parked; no driver is in the vehicle.
▶ quasi-dynamic WPT: Vehicle is stopped; driver is in the vehicle.
▶ dynamic WPT: Vehicle is in motion.

> **Definition**
> WPT: wireless power transfer.

There are also three WPT power classes (SAE J2954):

▶ light-duty home: 3.6 kW
▶ light _duty fast charge: 19.2 kW
▶ high-duty: 200–250 kW

With stationary charging, the electric energy is transferred to a parked vehicle (typically without passengers on board). It is important to keep the geometrical alignment of primary and secondary within certain tolerance values in order to ensure a sufficient efficiency rate of the energy transfer.

With quasi-dynamic wireless charging, the energy is transferred from the roadside primary coil system of limited length to the secondary coil of a slowly moving, or in stop-and-go mode moving vehicle (with passengers).

With dynamic wireless charging, the energy is transferred via a special driving lane equipped with a primary coil system at a high power level to a secondary coil of a vehicle moving with medium to high velocity.

7.2.2 **Stationary WPT**

In this option, electric vehicles simply park over an induction pad and charging commences automatically. WPT requires no charging poles or associated cabling. It can accommodate differing rates of charge from a single on-board unit, and the rate of charge or required tariff can be set from within the vehicle. It has no visible wires or connections and only requires a charging pad buried in the pavement and a pad integrated onto the vehicle.

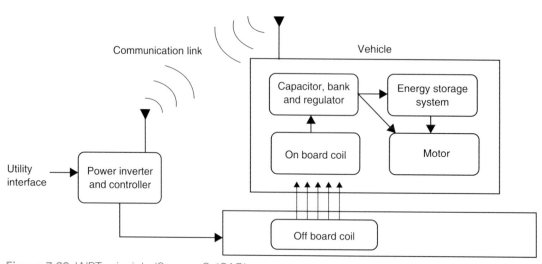

Figure 7.23 WPT principle (Source: CuiCAR)

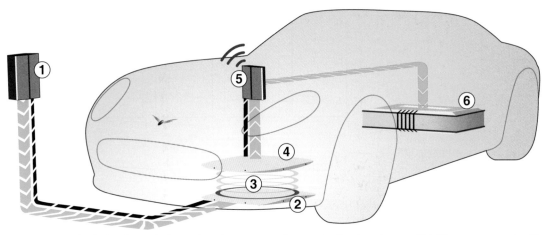

Figure 7.24 An inductive wireless charging system for statically charging an EV: 1) power supply, 2) transmitter pad, 3) wireless electricity and data transfer, 4) receiver pad, 5) system controller, 6) battery (Source: haloIPT)

> **Key Fact**
> Electric vehicles simply park over an induction pad and charging commences automatically.

The system works in a range of adverse environments including extremes of temperature, while submerged in water or covered in ice and snow. It will operate under asphalt or embedded in concrete and is also unaffected by dust or harsh chemicals. WPT systems can be configured to power all road-based vehicles from small city cars to heavy-goods vehicles and buses.

The power supply takes electrical power from the mains supply and energises a lumped coil, with a current typically in the range 5–125 A.

> **Definition**
> Power factor: the ratio of the real power flowing to the load to the apparent power in the circuit, expressed as a percentage or a number between 0 and 1.
> Real power: capacity of the circuit for performing work in a particular time.
> Apparent power: the product of the current and voltage of the circuit.

Pick-up coils are magnetically coupled to the primary coil. Power transfer is achieved by tuning the pick-up coil to the operating frequency of the primary coil with a series or parallel capacitor. The power transfer is controllable with a switch-mode controller.

A block diagram for a single-phase wireless charger is shown as Figure 7.18. The mains supply is rectified with a full bridge rectifier followed by a small DC capacitor. Keeping this capacitor small helps the overall power factor and allows the system to have a fast start-up with a minimal current surge. The inverter consists of an H-bridge to energise the tuned primary pad with current at 20 kHz. The 20 kHz current also has a 100 Hz/120 Hz envelope as a result of the small DC bus capacitor. Power is coupled to the secondary tuned pad. This is then rectified and controlled to a DC output voltage appropriate to the vehicle and its batteries. The conversion from AC to DC and back to AC, in the power supply side, is necessary so the frequency can be changed.

> **Definition**
> Inverter: an electrical device that converts direct current into alternating current.

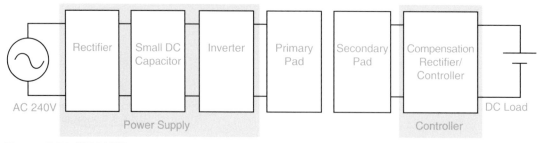

Figure 7.25 IPT (WPT) system components (Source: haloIPT)

The system includes three distinct hardware components:

1. High-frequency generator or power supply
2. Magnetic coupling system or transmitter/receiver pads
3. Pick-up controller/compensation

The high-frequency generator takes a mains voltage input (240 V AC at 50/60 Hz) and produces high-frequency current (>20 kHz). The output current is controlled and the generator may be operated without a load. The efficiency of the generator is high at over 94% at 2 kW. The generator comprises the following

▶ mains filter (to reduce EMI)
▶ rectifier
▶ bridge (MOSFETs) converting DC to high frequency
▶ combined isolating transformer/AC inductor
▶ tuning capacitors (specified for frequency and output current)
▶ control electronics (microcontroller, digital logic, feedback and protection circuits)

Definition

EMI: electromagnetic interference.

The design and construction of the transmitter and receiver pads gives important improvements over older pad topologies. This results in better coupling, lower weight and a smaller footprint, for a given power transfer and separation. The pads can couple power over gaps of up to 400 mm. The coupling circuits are tuned through the addition of compensation capacitors.

Safety First

All high-voltages are completely isolated, but note that safe working methods are necessary.

A pick-up controller takes power from the receiver pad and provides a controlled output to the batteries, typically ranging from 250 V to 400 V DC. The controller is required to provide an output that remains independent of the load and the separation between pads. Without a controller, the voltage would rise as the gap decreased and fall as the load current increased.

Key Fact

A CAN interface is used for control and feedback.

BMW has developed a wireless charging system. The principal benefit of this system is ease of use because drivers don't need to plug in a charging cable. As soon as the vehicle has been parked in the correct position above the inductive charging station, followed by a simple push of the Stop/Start button, the charging process is initiated. Once the battery is fully charged, the system switches off automatically.

Figure 7.26 Charging station (Source: BMW Media)

Energy from a main power supply is transferred to a vehicle's high-voltage battery without any cables. This can be installed in the garage. The BMW system also helps the driver to manoeuvre the car into the correct parking position with the help of a Wi-Fi connection between the charging station and vehicle. An overhead view of the car and its surroundings is displayed in the centre control display. Coloured lines help guide the driver into the correct spot. A graphic icon shows when the correct parking position for inductive charging has been reached. This can deviate from the optimum position by up to 7 cm longitudinally and up to 14 cm laterally.

The system consists of an inductive charging station (GroundPad) and a secondary vehicle component (CarPad) fixed to the underside of the vehicle. The contactless transfer of energy between the GroundPad and CarPad is conducted over a distance of around 7.5 cm. The GroundPad generates a magnetic field.

This induces a voltage in the CarPad, which then causes a current flow and charges the high-voltage battery.

During charging, ambient electromagnetic radiation is limited to the vehicle undercarriage. The GroundPad is permanently monitored and will be switched off if any foreign matter is detected. Foreign object detection and living object detection are part of the certified induction charging system. It will switch off charging if something is detected within the gap between the vehicle and ground pads.

The system has a charging power of 3.7 kW, enabling the high-voltage batteries on board a BMW 530e iPerformance, for example, to be fully charged in around three-and-a-half hours. The system has an efficiency of around 85%.

7.2.3 Dynamic WPT

It seems illogical in many ways but the prospect of wirelessly charging an EV as it

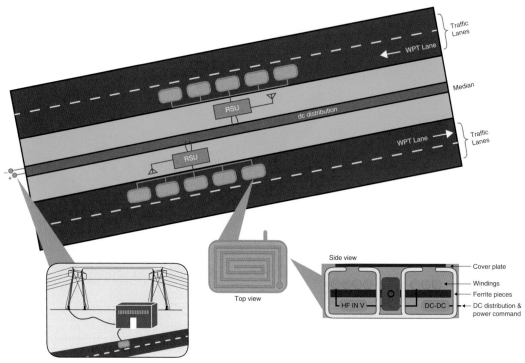

Figure 7.27 Principle of dynamic wireless charging – RSU is a roadside unit (Source: Oakridge National Lab)

drives along a road is already possible and being trialled in a number of countries. The principle is fundamentally the same as static wireless charging but even more complex. The technology is known as wireless power transfer (WPT).

The challenges with this technology are

▶ synchronisation of energising coils (timing of power transfer)
▶ acceptable power levels
▶ vehicle alignment
▶ allowable speed profiles
▶ multiple vehicles on charging lane

A number of feasibility studies and trials are ongoing, and it is expected that this system will be available in the near future. Figure 7.27 shows the principle of dynamic WPT.

Driver assistance systems may play a role in combination with wireless charging. With stationary wireless charging, a system could be developed where the vehicle is parked automatically, and, at the same time, primary and secondary coils are brought into perfect alignment. With quasi-dynamic and dynamic charging, the vehicle speed as well as horizontal and vertical alignment could be automatically adapted by dynamic cruise control and lane assist. This would increase the efficiency rate of the energy transfer because of the need to synchronise energy transfer via the coil systems.

Key Fact

Driver assistance systems may play a role in combination with wireless charging.

Communication will be essential to exchange standardised control commands in real time between the grid and the vehicle control systems.

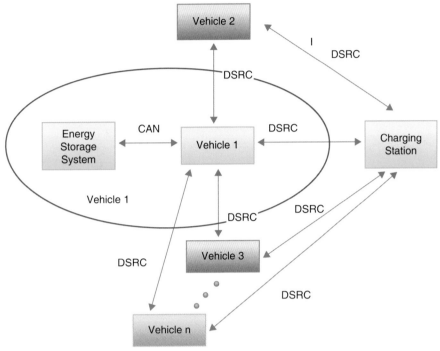

Figure 7.28 Communication is essential for dynamic WPT

For safety reasons, vehicles in other lanes and other users of the main coil system in the charging lane need to be monitored in real time.

Dedicated short-range communication (IEEE 802.11p) is a likely technology that will be used for low-latency wireless communication in this context.

Finally, powerful electromagnetic fields are used in the active charging zone between primary and secondary coils. For human safety in this respect, compliance with international standards needs to be achieved.

Note

1 https://www.gov.uk/government/statistics/ electric-vehicle-charging-device-statistics-july-2023/electric-vehicle-charging-device-statistics-july-2023

CHAPTER 8

Maintenance, repairs and replacement

8.1 Before work commences

8.1.1 Introduction

Before carrying out any practical work on an EV, you should be trained or supervised by a qualified person. Refer to Chapter 1 for procedures on safe working and PPE and making the system safe for work. Key aspects of practical work on EVs (and any other vehicles for that matter) are as follows:

- ▶ observation of health and safety
- ▶ correct use of PPE
- ▶ correct use of tools and equipment
- ▶ following repair procedures
- ▶ following workplace procedures
- ▶ referral to manufacturer-specific information

> **Safety First**
> Before carrying out any practical work on an EV, you should be trained or supervised by a qualified person.

Note: There are some excellent training courses available across the UK relating to electric vehicles and other subject areas. The IMI offer a range of CPD courses, and you can search here:

- ▶ The Institute of the Motor Industry: www.theimi.org.uk/cpd/search

I am pleased to recommend the theory and hands-on practical courses run by:

- ▶ ProMoto: www.pro-moto.co.uk
- ▶ TechTopics: www.techtopics.co.uk

8.1.2 Technical information

There are many sources of technical and other information, but in the case of EVs, it is essential to refer to the manufacturer's data, safety data sheets and workshop manuals. You should also be aware of the appropriate technique for gathering information from drivers/customers. For example, asking questions politely about when, where or how a problem developed can often start you on the road to the solution. General sources of information include the following:

- ▶ paper-based, electronic
- ▶ on vehicle data/warnings
- ▶ wiring diagrams
- ▶ repair instructions
- ▶ bulletins
- ▶ verbal instruction

An example of a manufacturer's data about component location is shown in Figure 8.1.

DOI: 10.1201/9781003431732-8

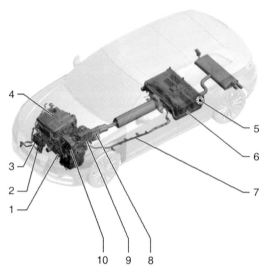

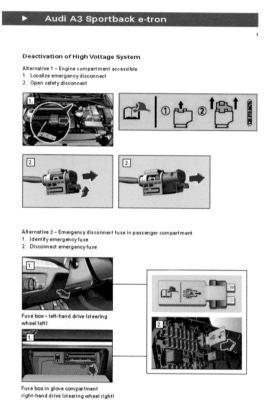

Information from rescue data sheets is also very useful. An example from Audi is shown as Figure 8.2.

Figure 8.1 Components on a Volkswagen Golf GTE: 1) three-phase current drive (electric drive motor, drive motor temperature sensor, 2) high-voltage battery charging socket, 3) electrical air conditioner compressor, 4) combustion engine, 5) battery regulation control unit, 6) high-voltage battery, 7) high-voltage cables, 8) high-voltage heater (PTC), 9) power and control electronics for electric drive (control unit for electric drive, intermediate circuit capacitor, voltage converter, DC/AC converter for drive motor), 10) charging unit 1 for high-voltage battery (Source: Volkswagen Group)

Figure 8.2 Sample page from the Audi rescue data sheets (Source: Audi)

Figure 8.3 Sample from the Pro-Moto app

A range of apps are now also available that include useful information, but remember, nothing beats the manufacturer, so this is always the data you should use where possible.

8.1.3 De-energising

Different manufacturers have different ways to de-energise the high-voltage system (you must refer to specific data for this operation). Here I have presented a typical but generic example of a de-energisation process:

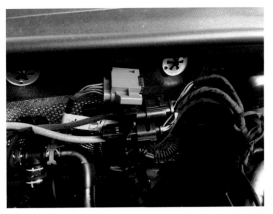

Figure 8.4 Maintenance connector (green) and warning label on a Golf GTE

> **Safety First**
> Always refer to the manufacturer's data before de-energising the high-voltage system.

Here is a typical routine:

1. Use appropriate PPE at all times.
2. Place high-voltage warning signs and a fence around the vehicle with posts and barrier tape or similar.
3. Disconnect the charging plug.
4. Switch ignition ON, connect scanner and check for faults.
5. Connect scanner and check that high-voltage readings are normal.
6. Switch OFF the ignition.
7. Remove the service connector.
8. Lock service connector to prevent accidental re-connection.
9. Switch ON the ignition.
10. Check dashboard warnings.
11. Connect scanner and check that high-voltage readings are ZERO
12. Switch OFF the ignition and remove the key to a safe place (at a distance if a remote key)
13. Check correct operation of a multimeter on a low-voltage source
14. Check for ZERO voltage at the inverter using a Cat III (minimum) multimeter and leads.

In some cases, a technician qualified in high-voltage may carry out the de-energisation process for others to then do specific work.

Figure 8.5 Warning signs used when work is being carried out on a vehicle after de-energisation

8.2 During work

8.2.1 Maintenance intervals

Maintenance intervals on electric and hybrid vehicles vary between models and manufacturers. Always refer to the manufacturer's data. Using an eGolf as an example, inspections depend on the length of time and the mileage. The first inspection takes place after 30,000 km or 24 months, followed by every 12 months or 30,000 km after that, depending on which comes first. The brake fluid inspection is recommended after three years, followed by every two years.

187

8.2.2 Repairs affecting other vehicle systems

If care is not taken, work on any one system can affect another. For example, removing and replacing something as simple as an oil filter can affect other systems if you disconnect the oil pressure switch accidently or break it by slipping with a strap wrench.

For this reason – and, of course, it is even more important when dealing with a high-energy system – you should always consider connections to other systems. A good example would be if disconnecting a 12 V battery on an HEV, the system could still be powered by the high-voltage battery and DC–DC converter. Manufacturer's information is, therefore, essential.

Electromagnetic radiation (or interference) can affect delicate electronics. Most engine control units are shielded in some way, but the very high–strength magnets in the rotor of some EV motors could cause damage. From a different perspective, an interesting news article recently noted how lots of drivers in a specific car park could not unlock their vehicles remotely.[1] This is still under investigation, but it sounds like an electromagnetic radiation (EMR) issue to me – power lines nearby perhaps!

> **Definition**
> EMR: electromagnetic radiation.

8.2.3 Inspect high-voltage components

During any service or repair operation, it is important to inspect high-voltage components. This includes the charging cables shown as Figure 8.6.

8.2.4 Cables

Two key aspects when inspecting high-voltage components to be aware of are

Figure 8.6 Domestic mains and charge point cables

- ▶ the current draw capability of the vehicle
- ▶ potential for short circuits and the subsequent vehicle component damage

In addition, you must be able to identify components and the connection methods used. High-voltage components and high-voltage cables should undergo a visual check for damage and correct routing as well as security. Pay attention to the following during the visual check:

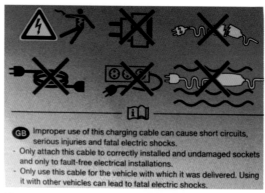

Figure 8.7 Warning and information supplied as part of the charging cables

- ▶ any external damage to the high-voltage components
- ▶ defective or damaged insulation of high-voltage cables
- ▶ any unusual deformations of high-voltage cables

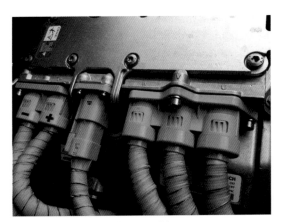

Figure 8.8 High-voltage cables

8.2.5 **Battery**

Check high-voltage batteries for

- ▷ cracks in upper part of battery housing or battery tray
- ▷ deformation of upper part of battery housing or battery tray
- ▷ colour changes due to temperature and tarnishing of housing
- ▷ escaping electrolyte
- ▷ damage to high-voltage contacts
- ▷ fitted and legible information stickers
- ▷ fitted potential equalisation line
- ▷ corrosion damage

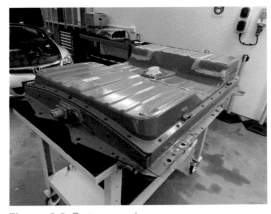

Figure 8.9 Battery pack

8.2.6 **Other components**

Engine compartment area: Check the condition of the power and control electronics

for electric drive, high-voltage cables for battery and air conditioning compressor, high-voltage cable for electric drive as well as high-voltage charging socket in radiator grille or in tank cap as appropriate.

Underbody: Check high-voltage battery as well as high-voltage cables for battery.

Figure 8.10 Battery pack bonding strap on a Nissan LEAF

8.3 **Remove and replace**

8.3.1 **High-voltage components**

The main high-voltage (also described as high-energy) components are generally classified as

- ▷ cables (orange!)
- ▷ drive motor/generator

Figure 8.11 The AC compressor on this car is driven by a high-voltage motor but still deep down on the side of the engine. You can just make out the orange supply cable here!

189

Figure 8.12 Toyota Prius engine bay

Figure 8.13 Volkswagen Golf GTE engine bay

- battery management unit
- power and control unit (includes inverter)
- charging unit
- power steering motor
- electric heater
- air-conditioning pump

General safety points to note with this type of work (see also Chapter 2, "Safe Working, Tools and Hazard Management"):

- Poisonous dust and fluids pose a health hazard.
- Never work on high-voltage batteries that have short-circuited.
- Danger of burns from hot high-voltage battery is a possibility.
- Hands may sustain burns.
- Wear protective gloves.
- Cooling system is under pressure when the engine is hot.
- Risk of scalding to skin and body parts is present.

- Wear eye protection.
- Risk of severe or fatal injury due to electric shock is present.

All high-voltage remove-and-replace jobs will start with the de-energising process and, after completion, the re-energising process.

Manufacturer's information is essential for any remove-and-replace job that involves high-voltage. Generic instructions for any component would be something like the following but more detailed:

1. De-energise the system.
2. Drain coolant if appropriate (many high-voltage components require cooling).
3. Remove any covers or cowling.
4. Remove high-voltage cable connections (for safety reasons, some connectors are double locked, Figure 8.14 shows an example).

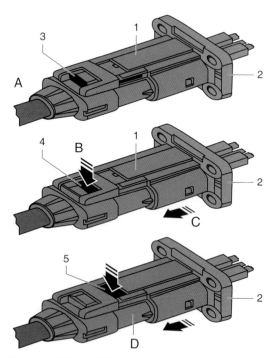

Figure 8.14 Locking connectors: Pull out lock 3 in direction A; Push mechanism 4 in direction B and pull off connector 1 until it is against the second lock; Push mechanism 5 in direction D and the connector can now be removed completely

5. Remove securing bolts/nuts as necessary.
6. Remove the main component.

Safety First
Manufacturer's information is essential for any remove-and-replace job that involves high-voltage.

8.3.2 Battery pack

For most battery-removal jobs, special tools and workshop equipment may be necessary, for example:

▶ hose clamps
▶ scissor-type lift platform
▶ drip tray
▶ protective cap for power plug

Typical removal process:

1. De-energise high-voltage system.
2. Remove underbody covers.
3. Remove silencer.
4. Remove heat shield for high-voltage battery.
5. Open filler cap on coolant expansion tank.
6. Set drip tray underneath.
7. Remove potential equalisation line.
8. Disconnect high-voltage cables.
9. Fit protective cap onto high-voltage connection.
10. Clamp off coolant hoses with hose clamps.
11. Lift retaining clips, remove coolant hoses from high-voltage battery and drain coolant.
12. Prepare scissor-type lift assembly platform with supports.
13. Raise lift assembly platform to support the high-voltage battery.
14. Remove mounting bolts.
15. Lower high-voltage battery using lift assembly platform.

Installation is carried out in the reverse order; note the following:

▶ Tighten all bolts to specified torque.
▶ Before connecting high-voltage cable, pull protective cap off high-voltage connection.
▶ Refill coolant.
▶ Re-energise high-voltage system.

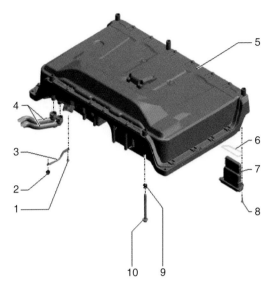

Figure 8.15 Battery pack: 1) bolt, 2) nut, 3) potential equalisation line, 4) coolant hoses, 5) high-voltage battery, 6) gasket, 7) battery regulation control unit, 8) bolt, 9) captive nut, 10) bolt (Source: Volkswagen Group)

8.3.3 Low-voltage components

As well as high-voltage, a large part of our work on EVs will be working on the low-voltage systems. Sometimes these are described as 'low-energy' to distinguish them from the 'high-energy' components such as the drive motor – but do remember that components such as a starter motor are low-voltage but not low energy! Low-voltage systems will include

▶ control units/fuse boxes
▶ low-energy components associated with interior heating
▶ wiring harness/cabling
▶ battery
▶ starter motor
▶ alternator
▶ switches
▶ lighting
▶ low-energy components associated with air conditioning
▶ alarm/immobiliser
▶ central locking

Figure 8.16 Standard 12 V starter motor (Source: Bosch Media)

▶ electric windows/wipers/washers
▶ central locking

For full details on these systems, please refer to one of our other books: *Automobile Electrical and Electronic Systems* (www.tomdenton.org).

8.4 Completion of work

8.4.1 Re-energising

Different manufacturers have different ways to re-energise the high-voltage system (you must refer to specific data for this operation). Here I have presented a typical generic example of a re-energisation process:

> **Safety First**
>
> Different manufacturers have different ways to re-energise the high-voltage system – you *must* refer to specific data for this operation.

1. Use appropriate PPE at all times.
2. Unlock service connector.
3. Refit the service connector.
4. Switch ON the ignition.
5. Check dashboard warnings.
6. Connect scanner and check for faults.
7. Remove the fencing and high-voltage warning signs.

8.4.2 Results, records and recommendations

This alliterative section is often overlooked, but it is very important to make a final check of any test results, keep a record of them and then, when appropriate, make recommendations to the customer. To interpret results, good sources of information are essential. All manufacturers now have online access to this type of information. It is essential that proper documentation is used and that records are kept of the work carried out. For example (usually electronic version),

▶ job cards
▶ stores and parts records
▶ manufacturer's warranty systems

These are needed to ensure the customer's bill is accurate and also so that information is kept on file in case future work is required or warranty claims are made. Recommendations may also be made to the customer, such as the need for

▶ further investigation and repairs
▶ replacement of parts

Or, of course, the message the customer would like to hear, that no further action is required!

Figure 8.17 Electronic scanner and data source

Recommendations to your company are also useful, for example, to improve working methods or processes to make future work easier or quicker.

Results of any tests carried out will be recorded in a number of different ways. The actual method will depend on what test equipment was used. Some equipment will produce a printout, for example. However, results of all other tests should be recorded on the job card. In most cases this will be done electronically, but it is the same principle.

> **Key Fact**
>
> Always make sure that the records are clear and easy to understand.

8.5 Recovery and roadside assistance

8.5.1 Introduction

Some EVs require special handling when it comes to recovery and roadside assistance operations. EV manufacturers provide detailed information and a number of other sources are becoming available such as phone apps.

> **Safety First**
>
> Some EVs require special handling when it comes to roadside assistance and recovery operations.

Much of the information in these sections is freely available to first responders and is provided by Tesla Motors on its website (www.teslamotors.com/firstresponders).

8.5.2 Emergency start

On some vehicles, even if the high-voltage battery has been discharged completely, there is still an option allowing the car to be restarted twice for a short distance. The eGolf, for example, has an emergency start function:

1. For approx. 100 metres, after switching the ignition off and on
2. For approx. 50 metres, after switching the ignition off and on once again

No further emergency starts are possible.

8.5.3 Roadside repairs

Roadside repairs should only be carried out by qualified personnel and by following all the safety and repair procedures outlined

Figure 8.18 Volkswagen eGolf on charge (Source: Volkswagen Media)

previously in this book. General information as well as more specific details are available.

8.5.4 Recovery

For roadside recovery, many manufacturers provide roadside assistance numbers for the driver to call. In addition, detailed data sheets are provided that give information similar to the following instructions for transporters (provided by Tesla in relation to the Model S):

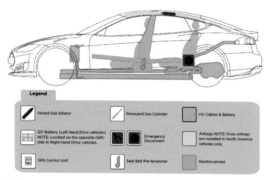

Figure 8.19 Key component and high-voltage information (Source: Tesla Motors)

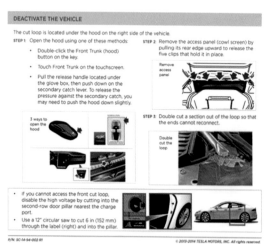

Figure 8.20 General instructions and deactivation information (Source: Tesla Motors)

8.5.5 Towing

Electric vehicles have a fixed connection between the drive wheels and the three-

phase current drive (electric drive motor/generator). This connection cannot be undone without mechanical work. If the vehicle needs to be towed, there are usually two options (For example, these instructions for the Volkswagen eGolf):

1. Towing the vehicle with the high-voltage system intact: Switch the ignition On and engage the selector lever in the N position to allow electric freewheel mode. The vehicle can now be towed for a maximum distance of 50 km at 50 km/h using a rope or towbar. Using a bar for towing is recommended for safety reasons.

2. Towing with a damaged high-voltage system: If it is not possible to activate the high-voltage system, the vehicle must be transported with all four wheels stationary. Freewheel mode cannot be activated, as there is a risk of overheating.

As a general guide, tow an EV slowly for a short distance to remove it and yourself from danger. Then suspend or use a flat bed.

8.5.6 Emergency response

In an earlier section, I outlined some general aspects that are important for emergency personnel. All manufacturers supply detailed information:

▶ model identification
▶ high-voltage components
▶ low-voltage system
▶ disabling high-voltage
▶ stabilising the vehicle
▶ airbags and SRS
▶ reinforcements
▶ no-cut zones
▶ rescue operations
▶ lifting
▶ opening

Key Fact

Most manufacturers supply detailed information.

A good example of this material is available from Tesla Motors on its website: www.teslamotors.com/firstresponders

8.5.7 Fire

A fire involving a BEV or HEV should generally be approached in the same manner as a conventional motor vehicle, although several additional factors should be considered. One approach indicating the basic steps that should be considered for extinguishing a fire involving any motor vehicle (including an EV or HEV) is illustrated in Figure 8.21:

Vehicle extrication and rescue and/or vehicle fire involves key steps to stabilise and disable the vehicle. A vehicle may appear to be off, but there may still be a hazard for fire fighters. Emergency responders must always disable the vehicle's ability to operate as per the vehicle manufacturer's instructions.

Emergency response guides from vehicle manufacturers usually recommend a defensive approach to a battery fire. In other words, let it burn and consume itself. However, exposure to the heat and/or products of combustion must be considered.

Methods for extinguishing a vehicle battery fire depend on many things such as type of battery, extent of fire, configuration and physical damage to battery unit (Figure 8.21). If water is used, copious amounts are normally required. However, this may be impractical if the vehicle and battery unit are not accessible and/or runoff is a concern. Some emergency organisations advise that for lithium-ion battery fires, extinguishment can be attempted using dry chemical, CO_2, water spray or regular foam. High-voltage batteries are well-sealed and do not contain much liquid electrolyte. Most spills can, therefore, normally be handled

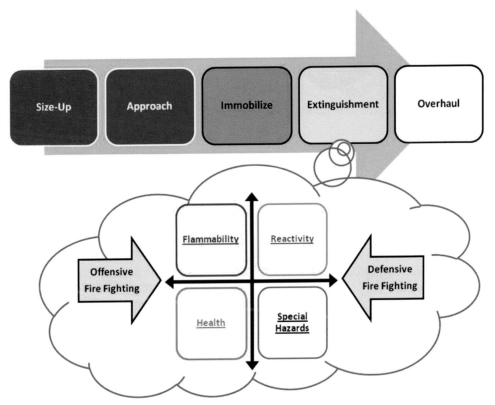

Figure 8.21 Example of approach to vehicle fire extinguishment

Figure 8.22 Fire blanket to cover an entire car (Source: Eintac)

with an absorbent. As new battery designs and technologies are introduced, this basic approach may need to be re-evaluated.

A further consideration is that built-in protection measures, which prevent electrocution from the high-voltage system, may be prevented from working properly. The normally open relays (contactors) for the high-voltage system could fail in a closed position if exposed to heat or if they are damaged. In serious accidents, short circuits to the chassis/body may become possible with the energy still contained in the high-voltage battery or any of the

high-voltage components. Always plan for the worst-case scenario. Figure 8.22 shows a fire blanket that can be used to completely cover a vehicle. Note this would control a fire but may not extinguish a battery fire.

If a fire starts on an EV (or any vehicle), the fire brigade should be called immediately and normal fire procedures followed.

EV specific guidance: If the vehicle is plugged in for charging, and it is safe to do so, it should be disconnected or the power switched off. Most fires on an EV should then be treated just like on an ICE vehicle.

▶ The normal range of extinguishers can be used if it is safe to do so.
▶ The vehicle (EV or ICE) can be covered with a fire blanket if it is safe to do so.
▶ If a fire starts on an EV battery that has been removed from the vehicle, it can be covered with a battery fire blanket if it is safe to do so.

If any of these steps have been carried out, then this should be communicated to the fire and emergency services.

8.5.8 Damaged vehicle delivery

An accident-damaged vehicle brought to the workshop should be

▶ **immediately marked with a red roof-mounted warning cone**
▶ de-energised or checked for safety by a Level 3 qualified technician, using suitable PPE
▶ offloaded and stored in the designated secure area
▶ left for a number of days as a safety precaution before any further work is carried out

There is a small risk that an EV battery could leak electrolyte, particularly if damaged. A dedicated spill kit must be provided and the instructions it contains followed if a spill occurs (Figure 8.23).

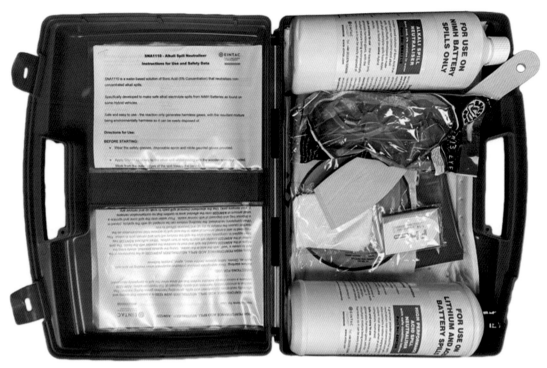

Figure 8.23 Battery leakage kit (Source: Eintac)

8.5.9 Accident and incident reporting

All accidents and incidents involving hazardous voltages and/or hazardous substances must be reported via the accident reporting system. For all exposure incidents/accidents or dangerous occurrences which have (or could have) resulted in death, injury or occupational disease, the health and safety executive will need to be notified through the 'Reporting of Injuries' process.

8.6 Transportation and storage

8.6.1 Electric vehicles

Undamaged EVs should be treated like any other vehicle on the premises. If an EV is damaged, then it should be treated as outlined previously. EVs should only be transported on a flatbed truck.

8.6.2 High magnetic strength components (motors)

These components should be handled like any other heavy item, but this should NOT be done by any person with a heart pacemaker or other medical device that may be affected by magnetism. The component should be in a fenced off area in the workshop or parts department. No additional precautions are needed. The old components should be packaged in the same way that the new one was delivered.

Transportation is by the normal methods for new and returned parts.

8.6.3 High-voltage batteries

When new and still packaged, these components should be handled like any other heavy item. Old or unpacked new batteries, modules or cells should be in a fenced off area

197

in the workshop or parts department. Only a qualified person should handle these items. The old components should be repackaged in the same boxes that new ones were delivered in, before being moved from the fenced off work area to the parts department or other designated storage area.

When being transported the original packaging must be used together with the appropriate warning label (UN3480 is for batteries only, UN3481 is used if the batteries are part of another assembly).

In most cases, a specialist shipping company should be used to transport batteries. This section is only intended as an overview of the issues.

New lithium-ion battery handling labels and shipping marks have been effective, under changes to the International Air Transport Association (IATA) dangerous goods transport regulations since 1 Jan 2017. However, under the 60th edition of the regulations (1 January 2019), labels such as those shown as Figures 8.24 and 8.25 are mandatory.

From 1 January 2020, manufacturers and subsequent distributors of lithium cells or batteries and equipment powered by lithium cells and batteries manufactured after 30

Figure 8.25 Mandatory label if the batteries are part of another assembly

June 2003 must make available the test summary as specified in the *UN Manual of Tests and Criteria, Revision and amend. 1, Part III, subsection*. Before lithium cells/batteries can be transported, they must have successfully passed certain tests. These tests simulate transport conditions like pressure, temperature, crush and impact and are described in the *UN Manual of Tests and Criteria*.

Types of packaging are also specified in regulations as outlined in Figure 8.26. It is usually recommended that a specialist company be used for transporting batteries – particularly if damaged.

Always refer to the latest information and regulations – this section is just intended as a guide.

8.6.4 Storage of cells and battery packs

To mitigate risks previously noted, individual cells or completed battery packs should be stored in a lockable steel shipping/storage container. Only qualified persons should have access to the storage area.

The container must be positioned as far away from the main buildings as possible. If

Figure 8.24 Mandatory label for batteries only

Lithium cell or battery test summary in accordance with sub-section 38.3 of Manual of Tests and Criteria
The following information shall be provided in this test summary:
(a) Name of cell, battery, or product manufacturer, as applicable;
(b) Cell, battery, or product manufacturer's contact information to include address, phone number, email address and website for more information;
(c) Name of the test laboratory to include address, phone number, email address and website for more information;
(d) A unique test report identification number;
(e) Date of test report;
(f) Description of cell or battery to include at a minimum:
(i) Lithium ion or lithium metal cell or battery;
(ii) Mass;
(iii) Watt-hour rating, or lithium content;
(iv) Physical description of the cell/battery; and
(v) Model numbers.
(g) List of tests conducted and results (i.e., pass/fail);
(h) Reference to assembled battery testing requirements, if applicable (i.e. 38.3.3 (f) and 38.3.3 (g));
(i) Reference to the revised edition of the Manual of Tests and Criteria used and to amendments thereto, if any; and
(j) Signature with name and title of signatory as an indication of the validity of information provided.

Figure 8.26 Lithium-ion packaging instructions taken from United Nations Economic Commission for Europe (UNECE) documentation

the worst should happen the resulting fire should be contained long enough for the fire and rescue service to arrive. The container(s) will be suitably signed to notify of the contents.

If it does become necessary to have any cells or battery packs stored inside during working hours, it is recommended that some or all of the following are observed:

▶ Insulated covers must be used.
▶ The items are positioned behind a sturdy barrier to prevent accidental damage from any workshop traffic (e.g., hand pump/pallet trucks or forklift trucks).
▶ Cells or battery packs shall never be stacked in a way that could cause the external structure to become damaged or compromised in any way.
▶ All 'live' cells or battery packs will be segregated and clearly labelled/signed as

such to adequately warn everyone in the facility of their presence.
▶ The items are only moved by a qualified member of staff who understand the associated risks.

8.7 Recycling industry guidance

Guidance has been produced for the recycling industry known as: *Electric and Hybrid Vehicles – Recyclers Best Management Practice Guide.*

Before vehicle collection

▶ Train drivers how to identify risks from these vehicles and processes required to make vehicles safe.
▶ PPE and neutralising kit should be available to drivers at all times and they should have received training in how to use them.

199

- Identify electric and hybrid vehicles prior to collection.
- Investigate type of incident and damage sustained (e.g., flood, fire, accident damage)
- Assess likelihood of battery damage (e.g., has vehicle sustained damage to area of vehicle where battery is housed) and warn driver of potential risks.

During vehicle collection

- Utilise PPE.
- Mark vehicle with appropriate warning signs.
- Assess vehicle condition and damage sustained prior to loading vehicle.
- Check battery status, including potential for any damage – inspect battery for physical damage, leakage or thermal incident (fire or discolouration of HV cables).
- Telephone for advice if necessary.
- Complete neutralisation of any spilt battery fluid.
- If necessary, de-energise vehicle, following manufacturer processes.
- If unsure – LEAVE IT WHERE IT IS and get a specialist.

Vehicle on-site

- Make suitable tools and PPE available.
- Access vehicle information resource (see examples at end of section).
- Reassess vehicle for potential HV battery or HV system damage – inspect battery for physical damage, leakage or thermal incident (fire or discolouration of HV cables).
- Mark vehicle with appropriate warning signs.
- De-energise vehicle, ALWAYS FOLLOW MANUFACTURER PROCESS – if unsure seek advice before de-energising.
- Once de-energised, store vehicle in a quarantined area, on flat ground or suitable racking.
- Train staff how to handle vehicle correctly to avoid any further damage to HV systems and allow only trained staff to work on vehicle until it is deemed safe.

- If flood-damaged, lift one end of the vehicle to drain as much water as possible prior to storage.
- Make vehicle watertight prior to storage.

Dismantling

- Make suitable tools and PPE available.
- Never work on vehicle alone – always have colleagues around 'just in case'.
- Manual handling guidelines must be followed.
- Follow manufacturer's guidelines or other resource (see examples at end of section) when dismantling vehicle.
- Allow only trained staff to work on vehicle until it is deemed safe.
- Store battery appropriately in a restricted access area – never mix battery types, store in the same orientation as when in the vehicle, store in such a manner that they cannot fall, or have anything fall on them.

Resources

- IDIS (www.idis2.com)
- NFPA (www.nfpa.org)
- EINTAC (https://eintac.com)
- Automotive Recyclers Association – ARA University
- individual vehicle manufacturers

Note: This is a guide and template only. All vehicle recyclers are advised to complete their own risk assessments and best management practices that are suitable for their specific requirements, operating practices, local rules, legislation, training and standards. This guide and lots more useful information relating to dismantling and recycling is available from Salvage Wire: www.salvagewire.com.

Note

1 https://www.ofcom.org.uk/news-centre/2021/spectrum-mystery-leaves-shoppers-stuck-in-car-park

CHAPTER 9

Diagnostics

9.1 Diagnostic operations

9.1.1 Introduction

This section is designed to not only give an overview of how to carry out this type of work but also what is actually possible. Note, you must do proper training and be qualified before carrying out any work on high-voltage systems. This is for your own safety as well as the vehicle's driver and passengers.

9.1.2 Live data

For the live data examples in this section, I used a TopDon ArtiPad or the later Phoenix version. More details are available from: www.diagnosticconnections.co.uk.

Scanning for live data on a vehicle is just as important as scanning for diagnostic trouble codes (DTCs). Usually, a DTC is the first step and then live data is the second. In this

---

Safety and information

Electric vehicles (EVs) have

▶ high-voltage components and cabling capable of delivering a fatal electric shock
▶ stored electrical energy with the potential to cause an explosion or fire
▶ components that may retain a dangerous voltage even when switched off
▶ the potential to affect medical devices such as insulin pumps and pacemakers

Before working on EVs, you must

▶ be qualified to the appropriate level (and ideally IMI TechSafe™ registered)
▶ refer to the vehicle manufacturer's information
▶ use all equipment in accordance with its safety instructions

IMI TechSafe™

The *Electricity at Work Regulations* (1989)[1] state that: 'No person shall be engaged in any work activity where technical knowledge or experience is necessary to prevent danger or,

DOI: 10.1201/9781003431732-9

> where appropriate, injury, unless he possesses such knowledge or experience, or is under such degree of supervision as may be appropriate having regard to the nature of the work'.
>
> IMI TechSafe™ registration covers this requirement and further ensures that complex automotive technologies are safely and correctly repaired by qualified technicians. To be registered as IMI TechSafe™, you must:
>
> ▶ successfully complete an appropriate qualification or IMI Accredited solution
> ▶ join the IMI Professional Register
> ▶ complete CPD from a specified range to maintain professional registration

section, we will look at a few examples of the information that is available relating to the high-voltage battery.

The screenshots shown as Figures 9.2 to 9.15 just scratch the surface of what information can be accessed from the vehicle. However, they serve to give a good overview of the process, whether the vehicle is ICE or EV or, in this case, both as it is a PHEV.

Safety First

You must do proper training and be qualified before carrying out any work on high-voltage systems.

Figure 9.2 This is the main menu screen where in this case I will choose the *Intelligent Diagnosis* option.

Figure 9.3 The scanner then connects to the Bluetooth vehicle connection interface (VCI) that I have already plugged into to car's data link connector (DLC). It then automatically scans for the VIN and, if found, sets the scanner up appropriately for the vehicle. In this case a 2018 Volkswagen Golf GTE. I then select the *Diagnostic* option.

Figure 9.1 Phoenix Pro scanner

Figure 9.4 The scanner now interrogates the network to see what is connected to it and displays a list of the various systems. I next select option *8C Battery Energy Control Module*.

Figure 9.5 It then takes a minute as the scanner interrogates the specific control system.

Figure 9.6 Now we have the options as displayed here. The two of interest for now are *02 Read DTC* and *08 Read Data Stream* (live data). I select *Read Data Stream*.

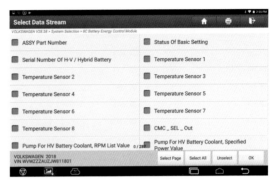

Figure 9.7 The screen now shows a very long list of the parameters that are being monitored and can be displayed as live data. For example, you can see here that there are eight temperature sensors on the battery pack. This data could be very useful if you were looking for a specific fault in the battery.

Figure 9.8 I have scrolled down the list here, and, for example, you could now select to view high-voltage battery current and voltage readings.

9.1.3 **Insulation testing**

An insulation tester is mostly used on electric and hybrid vehicles to check that the high-voltage cables and components are safe (see also Section 2.5.3).

The device shown as Figure 9.16 is known as a Megger and can supply up to 1000 V to test the resistance of insulation on a wire or component. A reading well in excess of 10 MΩ

Figure 9.9 Scrolling down further allows you to select *Charge State, Cell 01* to *Cell 96*. I have just selected four of them for now. Scrolling down once again allows you to select *Cell 01* to *Cell 96*. These are the voltages of each cell in the battery pack. I have again just selected four of them for now. I also selected *H-V Battery voltage* and temperature on the screen we looked at previously.

Figure 9.10 Pressing *OK* now displays the chosen live data. In this case, it is showing the battery voltage as 387.2 V and the state of charge (SOC) of Cells 01 and 02 as 93%. Cells 03 and 04 were the same. The symbols on the right allow you to plot the data as graphs. Alternatively, pressing *Record* will keep a record of the figures over time.

is what we would normally expect if the insulation were in good order – but as always, check manufacturer's recommendations.

The high-voltage is used because it puts the insulation under pressure and will show up

Figure 9.11 Graphing is very useful if you want to monitor how the live data changes over time or even as the vehicle is being driven.

Figure 9.12 Going back one step and scrolling down we can see the voltages of Cells 03 and 04 display to 3d decimal places. Both are 4.026 V in this example. Comparing cell voltages across all 96 of them is a great way to track down a fault. The figures on this car were all within a few tens of millivolts – so all is well!

faults that would not be apparent if you used an ordinary ohmmeter.

Key Fact

Insulation testers use a high-voltage because it puts the insulation under pressure to show up faults that would not be apparent with an ohmmeter.

Figure 9.13 After returning to a previous menu. I then scan for faults in the *Battery Energy Control Module*. None were present.

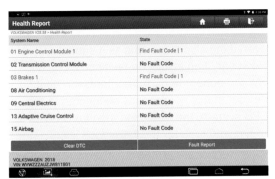

Figure 9.14 Finally, I scanned for DTCs (fault codes) across the whole vehicle. It is recommended that before any work is carried out on a vehicle that a full DTC scan is carried out. This is evidence of the state of the vehicle before you work on it. Doing another scan after work is complete shows you have repaired a fault and/or not created anymore. Four faults showed up.

To test the insulation resistance on the stator shown as Figure 9.17, the Megger should be connected between one of the three phase windings and the iron body.

Take care when using insulation testers. The high-voltage used for the test will not kill you because it cannot sustain a significant current flow, but it still hurts!

Figure 9.15 The four faults are all noted as passive faults so not serious enough to illuminate the warning light. More work for another day perhaps! Pressing the *Report* button will save this list so you can keep this information for your customer.

Figure 9.16 Megger insulation tester

205

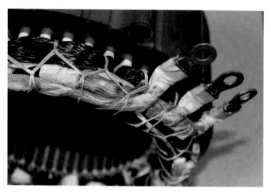

Figure 9.17 Stator from a hybrid motor

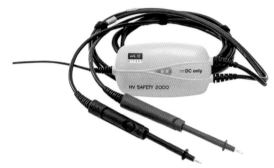

Figure 9.18 AVL HV Safety 2000

9.1.4 Equipotential bonding

In the UK aftermarket, it is not custom and practice to carry out an equipotential bonding (EPB) check, but in some EU states, Germany and Austria, for example, it is. Should we be doing an EPB check after EV repairs or service here in the UK? I will review the equipment later, but first let's learn more about EPB and how it relates to EVs.

EPB is the process of electrically connecting metalwork and conductive parts of various components so that the voltage is the same throughout these various parts (an equal potential). EPB is used to reduce the risk of equipment damage and personal injury. It is sometimes simply referred to as bonding and is commonplace in domestic and industrial electrical installations. Its main purpose is safety.

In the near future, all workshops and garages will have to deal with a significant number of high-voltage electric vehicles (EVs). Guaranteeing the safety and reliability of vehicle high-voltage systems is something we will need to become more familiar with. Training, qualifications and CPD to maintain registration will be essential.

EVs use an insulated return system such that there is no connection between the high-voltage circuits and the chassis. As part of the startup process (when the vehicle is switched into *Ready* mode), the on-board insulation monitor device is checked. Some

manufacturers also carry out the check frequently during vehicle's operation phase.

If a failure of this 'monitor' is detected, then the vehicle will not switch on – providing everything else is working as it should. However, if an insulation fault occurs on an HV component (e.g., AC compressor, heating device) that is not properly connected to the EPB system, the insulation monitor is not able to find and monitor this hazardous issue.

When repairs are carried out, the insulation resistance or isolation strength of high-voltage system components is always performed by the technician. The continuity of cables, including the braided shielding, is also tested. A value of 500 V is often used for this test, but for some systems, it is becoming necessary to use 1000 V, but this is not always available from some testers.

To date, here in the UK anyway, the bonding (equipotential bonding) resistance between the components and chassis has not been carried out as a matter of course – other than as a physical 'hand-and-eye' check.

The reason that all the components are bonded to the chassis is so that even if insulation faults occur, the potential (voltage) difference between all the main metal components will always be zero – so no electric shock will occur. In theory, this should never happen so the bonding is like a belt and braces approach – better safe than sorry though because the high-voltages used can be fatal. It is also

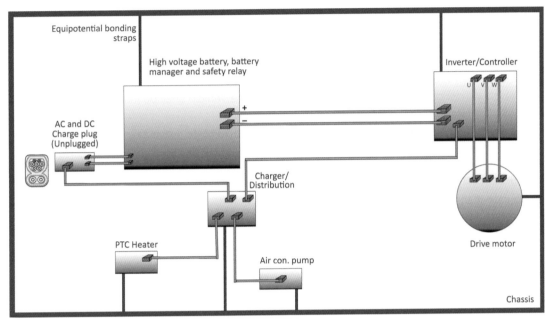

Figure 9.19 EV system block diagram showing main components and chassis equipotential bonding

important to note, as mentioned previously, a possible insulation fault can only be monitored if no EPB faults are present.

At this point, we should once again remind ourselves of Murphy's Law: *If it can go wrong, it will go wrong!*

It is becoming more common for technicians working on vehicles to be required to *prove*

that certain things have been carried out, for example, recalibration of the advanced drive assistance systems (ADAS) after a new steering component has been fitted. These checks or calibrations are now being specified as an insurance industry requirement (IIR) after major accident repairs have been completed.

Figure 9.20 This bonding strap connects the case of a high-voltage component to the chassis on a Volkswagen Golf GTE

Figure 9.21 A bonding strap between the battery box and chassis on a Nissan LEAF

A trained technician will always perform tests on a vehicle to make sure it is 'made safe' or 'de-energised' before any work is carried out that may involve the high-voltage system. This is different from when the question is asked by the vehicle owner or driver.

The vehicle runs through checks every time it is switched into *Ready* mode. A technician carries out more tests during or after repairs have been done, including electrical and physical checks. But how do you prove this has been done, particularly if something safety-related does go wrong with the HV system at a later date? Note that isolation strength (insulation resistance) of some components can deteriorate over time.

For some organisations, these sort of measurements and records are becoming more important. For individual technicians, it is also becoming important that they are guided through the tests so they can prove all tests have been carried out and prevent litigation.

The UN regulation ECE R100 is titled: *Uniform provisions concerning the approval of vehicles with regard to specific requirements for the electric power train*. There are two main parts to this in relation to high-voltage safety, as represented by Figure 9.22.

The specified equipotential bonding, low electrical resistance requirements are

1. The HV component and the vehicle body shall be less than 40 mΩ.

2. The HV component and an adjacent conductive part is less than 20 mΩ.
3. Between two HV components that are simultaneously accessible for a person and are arranged in a distance up to 2.5 m in the vehicle shall be less than 100 mΩ

The requirement that is important in the aftermarket (Number 3) is that the resistance between all exposed conductive parts and the chassis must be lower than 0.1 Ω (100 mΩ) when at least 200 mA is flowing.

Note that to measure milliohms (mΩ), special equipment is needed; a normal multimeter is not suitable.

EPB is used to reduce the risk of equipment damage and personal injury from high-voltages.

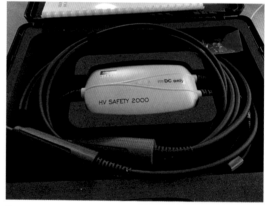

Figure 9.23 AVL DiTEST HV Safety 2000

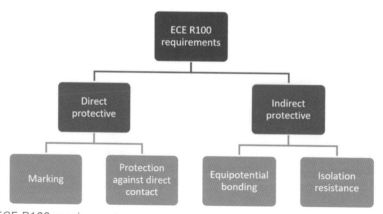

Figure 9.22 ECE R100 requirements

Checking EPD can be more involved than it may first appear. However, with the proper equipment, suitable training and time, it is very easy.

Checking that the EPB is meeting regulations is something that could become more common as more high-voltage vehicles hit our workshops. Measuring and recording EPB resistance will require a set process and suitable equipment. The AVL DiTEST HV safety equipment, a CAT III device, could be one solution.

This high-voltage measurement equipment guarantees the verification of all safety requirements for DC high-voltage (HV) circuits

on the vehicle. This computer-based test equipment is able to perform measurements according to UNECE R100. It supports the requirements of an EPB measurement with a current of 1 A. The equipment is powered from the USB port, no additional power supply or batteries are needed and no range selection is required, reducing the potential for errors.

After following ALL appropriate safety instructions, a fully trained EV technician can carry out a number of tests and is guided through the process using the equipment specified in Table 9.1 (after the initialisation/self-test):

Table 9.1 Features of the AVL DiTEST HV Safety 2000

All-pole DC voltage measurement	The absence of a voltage is verified. This is performed by a menu-driven measurement of the three relevant voltages: 1. Voltage measurement HV+ against HV- 2. Voltage measurement HV+ against the chassis 3. Voltage measurement HV- against the chassis
DC voltage measurement	This mode is used to measure any DC voltage (up to 1000 V)
HV insulation measurement	Tests for insulation resistance by applying a high-voltage of up to 1000 Vt
Equipotential bonding	The device generates a test current of up to 1 A. This current loads the bonding conductor accordingly so that faults can be detected. The test voltage generated for this is respectively low so that the measurement does not create a hazard. • Range: Measuring currents 100 to 1000 mA • Resolution: 1 mΩ • Tolerances: ± 3.5% v. M (of the measured value)
Resistance measurement	Resistors in the range of 1 mΩ to 10 MΩ can be measured
Diode test	Diodes can be tested in the forward direction and reverse direction. Open and short circuit states are determined and displayed
Capacity measurement	In this mode capacitors can be measured in the range 1 nF to 300 μF
Insulation monitor check	The function of the insulation monitor in a HV vehicle is tested and the trigger threshold determined. For this purpose, a series of voltage measurements between HV+ and the chassis are performed, whereby at each voltage measurement the internal resistance of the HV Safety 2000 is gradually decreased by steps
Insulation resistance	This function is used to measure insulation resistance in accordance with the recommendation of SAE J1766: www.sae.org/standards/content/j1766_201401/
Full measurement	This mode combines the following measurements into one user-guided procedure: 1. Prepare vehicle for the measurement 2. Verification of zero-potential 3. 3 HV Insulation measurement at the zero-potential system 4. SAE J1766 measurement on the live system

To use the system, the technician creates a profile, which can be password-protected. Before testing, the vehicle details are also entered. This is all saved and can be reused.

After each measurement, a detailed results protocol can be generated. By activating an *Automatic Save* function, a result protocol is generated and saved automatically after each measurement.

The documented results, which are dated, linked to the vehicle and technician and saved, are a key feature. This is robust proof that tests have been carried out by a specified person using calibrated equipment at a specific date and time.

The first step is to enter the technician's details or select yourself from the list if already done. Then the vehicle details are entered or, again, selected from a list of already done.

After completing technician and vehicle details, clicking the *Diagnostics* button shows the test/check menu.

The step-by-step guidance in the program was very easy to follow and ensured all stages of the tests/checks were covered, including calibration. Other than difficulty accessing HV+ and HV- on some vehicles, which, of course, is the case regardless of the equipment in use, all the tests were easy to carry out. On most vehicles, to access HV+/HV-, the inverter cover needs to be removed. Arguably, many of the checks could be done with a good multimeter and insulation tester, but, of course, this does not automatically keep a written record.

Not applicable in the UK at this time, but according to German and Austrian legislation, a multimeter is not allowed for this test because of the potential to adjust the voltage range or connect the leads incorrectly. A two-pole tester is recommended by some organisations. An exact copy of the SAE J1766- and ECE R100-based tests was not possible without the AVL equipment. The EPB testing was the most interesting check and is definitely not possible with a normal multimeter.

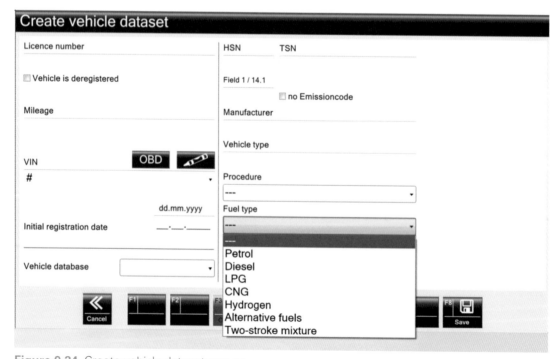

Figure 9.24 Create vehicle dataset screen

In principle, the test is very simple. A known current (up to 1 A as per ECE R100 requirements) is made to flow and the voltage is measured. The resistance is then calculated. Before this, if using standard leads, the system is calibrated or zeroed to ensure accuracy.

If the optional Kelvin leads are used, then no calibration is needed because the voltage at the test points is measured using a separate circuit to the one where the current is flowing. The Kelvin crocodile clips (Figure 9.25) have both jaws insulated from each other and are attached to a twin-core insulated cable.

One challenge, which is a little more complex than it first seems, was where to connect the test leads. To even see the battery pack on a Nissan LEAF, for example, a large protective cover has to be removed – it's not difficult, but it does take time. The pack is painted, and as this is an older car, so it is also a little corroded and dirty. We chose to connect one lead, as shown in Figure 9.26, after scratching a small amount of paint off (which will be replaced!). In this way the connection is to the actual pack rather than a bolt or strap. Polarity in the test is not important.

One option for the other lead was the bolt as shown in Figure 9.27, but our decision was that we couldn't be absolutely certain that this was connected to the chassis. It could just be contacting the strap.

In Figure 9.28 we cleaned the bolt and made the connection as shown. In this way, if the

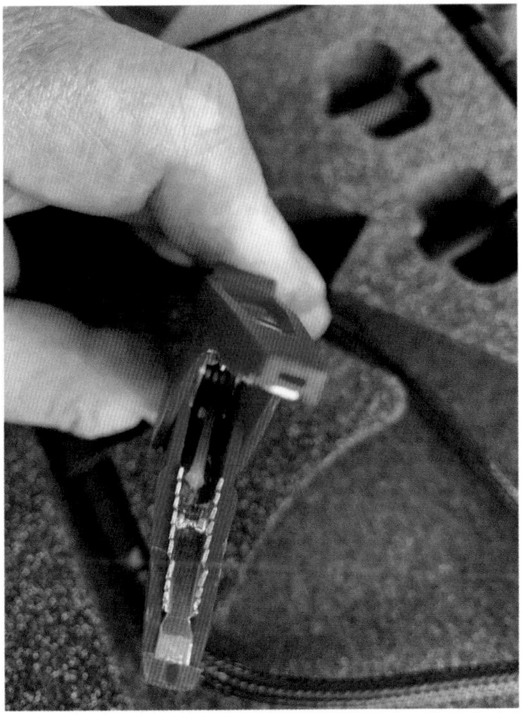

Figure 9.26 Kelvin probe connected to the battery pack

Figure 9.27 One option for connection that was NOT used (There are 4 of these bonding straps on the battery pack)

Figure 9.25 Kelvin lead crocodile clip showing insulated connections

Figure 9.28 Chassis bolt connection

result was as required (less than 100 mΩ) when we did the test, it would definitely prove a low-resistance circuit between the battery case and the vehicle chassis. Our result was 5 mΩ so well within specifications.

Checking the EPB on the motor of the vehicle shown in Figure 9.29 was easier to access and connect. One connection to the low-voltage battery negative terminal and one to the body of the motor after the area was cleaned. Again, we got a very low result of 8 mΩ.

Overall, after a little practice because the equipment was unfamiliar at first, the checks were very easy to carry out. The

Figure 9.29 Motor body connection on the Renault Zoe

automatically created report can be printed out, and an example is shown as Figure 9.30. You can see in this report that one test failed, but the next one passed. It is okay to try the test a few times if the result is a failure. This is because it is very easy to make a poor connection accidently but not possible to achieve a pass accidentally – so one good result means the resistance is within specs.

9.2 Diagnostic tools and equipment

9.2.1 Introduction

Diagnostic principles are the same for low or high-voltage vehicles, but there are additional issues that we must consider, not least of which is safety. We will look at

- ▶ tools and equipment (what is needed and examples of its use)
- ▶ non-invasive testing (what can be done without any dismantling)
- ▶ invasive and more in-depth testing (getting deeper into waveforms and how to interpret them)

9.2.2 De-energisation procedures

Before carrying out any work on an EV, it may be necessary to switch off the high-voltage system. This is described in different ways by different manufacturers (de-activation or isolation, for example). Manufacturers also have different ways to de-energise the high-voltage system so to be certain you are carrying out the process correctly, you must always refer to their specific data. Here I have presented a reminder of a typical but generic example of a de-energisation process:

1. Use appropriate PPE at all times and place high-voltage warning signs and a fence around the vehicle with posts and barrier tape.
2. Disconnect the charging plug.

Result protocol
HV Safety equipotential bonding measurement

04/02/2022 09:40

| | Phone: |
| | Fax: |

Licence plate:	MM65OZL	Manufacturer:	Renault
Mileage:	20000	Vehicle type:	Zoe
Vehicle identification number:		Engine code:	
Initial registration date:		Exhaust system:	---

Software

| Test software | HV Safety |
| Version | 2.5.8036.0 |

Device information

Device name	AVL HV Safety 2000
Serial number	5705
Firmware version	4.16
Production date	05/11/2021
Calibration date	05/11/2021
Calibrated by	5143
Calibration device	VAS5143A, SWB, Keithley2000
Zero adjustment accomplished	04/02/2022 09:37:56

Operator

| Operator | Tom |
| Safety notes | The operator accepted and confirmed the safety notices by pressing the Next button (F8) |

Measuring values

Measurement 1 - Equipotential bonding line resistance too high (> 0.1Ω)		
Measured on		04/02/2022 09:39:36
Test current	[mA]	1000
Resistance	[Ω]	↔0.303
Measurement 2 - Equipotential bonding line resistance ok		
Measured on		04/02/2022 09:40:40
Test current	[mA]	1000
Resistance	[Ω]	0.008

Figure 9.30 Example report showing that the first test failed

3. Switch ignition ON, connect scanner, check for faults and check high-voltage readings are normal.
4. Switch OFF the ignition.
5. Remove the service connector (this can be a low- or a high-voltage device).
6. Lock service connector to prevent accidental re-connection.
7. Switch ON the ignition and check dashboard warnings.
8. Connect scanner and check high-voltage readings are ZERO.
9. Switch OFF the ignition and remove the key to a safe place (at a distance if a remote key)
10. **Prove correct operation of a Cat III (minimum) multimeter or a two-pole tester on a low-voltage source, then check for ZERO voltage at the inverter and/or battery output.**

Step 10 of the generic de-energisation process is where the first and most important test is made.

Figure 9.31 Volkswagen service/maintenance connector breaks the pilot line, which opens the main contactors

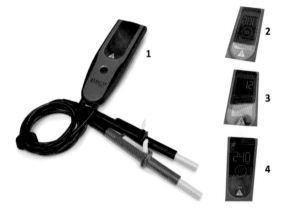

Figure 9.33 Two-pole tester and the four possible results (part of the PicoScope EV kit)

9.2.3 **Two-pole tester**

A voltage tester (Figure 9.33) has one job – it effectively says 'Go' or 'No-Go'. It could be the piece of equipment that takes the most important reading – the one that stops you being killed by high-voltage electric shock.

This device is deliberately very simple. It has one button to switch it on that also illuminates a handy light, and it will display one of only four possible results:

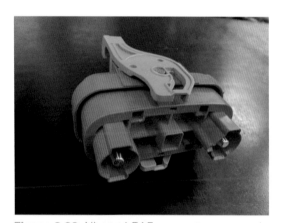

Figure 9.32 Nissan LEAF connector removed from a panel under the rear seat. This breaks the high-voltage battery circuit in the middle of the battery pack, effectively halving the high-voltage risk as well as isolating it from the rest of the car.

1. Leads open circuit – nothing is displayed.
2. Leads shorted together – zero is displayed in red and an alarm sounds (Test).
3. Leads connected to a voltage between 3 and 36 V – voltage value is shown as AC or DC and the display is blue (Go).
4. Leads connected to a high-voltage above 36 V – voltage value is shown as AC or DC, the display is red and an alarm sounds (No-Go).

Testing for a zero-volt potential requires a dedicated measurement routine to ensure beyond a reasonable doubt that the results are accurate and verified.

You can, of course, use a suitable CAT III multimeter to do this test. However, there is a potential for error because of the many settings and connections on a multimeter. For example, what happens if you apply a multimeter to the correct high-voltage test points of a vehicle, but you have set the dial to display AC voltage or the leads are in the wrong ports on the meter? In both cases, the multimeter could read either low or no voltage suggesting the vehicle is safe to work on – when it isn't!

9.2.4 **Multimeter and insulation tester**

Multimeters and insulation testers can be separate pieces of equipment, but many are

Figure 9.34 Simple low-voltage test (but importantly a check of the meter and leads)

Figure 9.36 Cheap but very useful amp clamp multimeter

Figure 9.35 Insulation resistance test between a charger connection and the vehicle body. The reading here is over 1000 MΩ so well within specifications

now combined into one device (Figures 9.34 and 9.35). Multimeters used for high-voltage work should be a *minimum of CAT III.*

9.2.5 Amp clamp

A simple and cheap device is shown as Figure 9.36. It is a CAT II multimeter, so it is not appropriate for EV testing. However, using the amp clamp is noncontact so no high-voltage reaches the device. The reading shown in Figure 9.36 is the current flowing from the high-voltage battery with the car stationary but

the heater switched on. More about what this reading means in the next sections.

9.2.6 Oscilloscope

There are several types of oscilloscope designed specifically for automotive diagnostics. My favourite by far is the PicoScope. Now even better, Pico has produced a kit specifically for high-voltage systems. The kit is based on the PicoScope 4425A with PicoBNC+® technology, which powers devices such as the amp clamps directly from the main unit – no batteries needed. EV guided tests, along with all the normal ICE systems, are available in the free PicoScope Automotive software.

Some examples of the EV guided tests are as follows:

▶ DC high-voltage energy system and the AC motor/generator drive system
▶ charging behaviour and 12 V/high-voltage battery current split issues
▶ CP/PP communications between charging system and vehicle
▶ three-phase current measurements to check winding balance under load

Some of these tests will be covered in the next sections. Because Pico, like the IMI, support the requirement that technicians are properly trained and qualified, the EV guided tests are

215

Figure 9.37 Top of the range PicoScope kit specially designed for EV diagnostics – note the three current clamps that can be used on each phase of the motor/generator circuit

only available in the dedicated PicoScope software.

9.2.7 **Differential probe**

An active differential probe allows high-voltage measurements, which an oscilloscope alone would not be suitable for. The probe shown is CAT III, rated up to 1000 Vrms and has a 25 MHz bandwidth. The voltage rating applies to its 1/20 and 1/200 (attenuation) settings.

This differential probe has an auxiliary grounding terminal. It is very important that this terminal is connected to a reliable ground

point on the vehicle to prevent transient voltages from damaging the scope.

Instructions for differential probes may imply that some oscilloscopes are grounded (earthed) through the mains connection. A USB oscilloscope connected to a laptop is unlikely to be grounded even if it is powered from the mains. However, for on-vehicle measurements, it is important to distinguish between earth as in the ground we walk on and chassis ground or earth on a vehicle. It is the vehicle ground that should be used for EV testing.

9.2.8 **Scanner**

As with ICE systems, a scanner is an essential piece of EV diagnostic equipment. There are two main parts to its abilities:

▶ reading diagnostic trouble codes (DTCs)
▶ displaying live data

DTCs identify a specific problem area and are a guide as to where a fault might be occurring within the vehicle. High-voltage system DTCs are displayed in the same way as any others. Likewise, live data can be shown on the screen, for example, battery voltage and even individual cell voltages in the traction battery.

In the next sections, we will look in more detail at some of the useful information that is available to assist with EV diagnostics.

Figure 9.38 Active differential probe

Figure 9.39 Scanner connected to the DLC using a Bluetooth dongle

9.2.9 **Summary**

There is still some debate in our industry about how many EVs (including HEVs, PHEVs and FCEVs) we will see in our workshops in the short term. What is certain is that in the medium and longer term, they will become very common. The right training and the right equipment for accurate and safe diagnostics will be essential.

9.3 **Non-invasive measurements**

9.3.1 **Introduction**

Taking a measurement and then reacting to it as part of a logical diagnostic routine is how to find faults in vehicle systems, EV or otherwise. In other words, test don't guess!

An aspect sometimes overlooked, however, is how the act of taking a measurement can change the value of the thing that you are measuring (known as the measurand). For example, a test lamp on a sensor circuit would load the circuit and change the values – or worse, damage the system. Even a multimeter set to measure voltage can load a circuit by drawing a small current, which changes the reading accordingly. This is known as an invasive or intrusive measurement, where the act of measuring changes the circuit so an incorrect reading is obtained. However, good quality multimeters have a very high internal resistance (10 MΩ or more) so have a negligible effect.

A non-invasive measurement is one where the instrument has no effect on what is being read. For example, a current clamp around a wire is one type of noncontact measurement. Another example is the coil-on-plug (COP) probe. Both of these are covered in more detail later.

9.3.2 **Known good waveforms and measurements**

'Known good' a phrase we often use when diagnosing faults. These measurements range from comparing the details of a known good waveform with the one you are currently viewing, to a voltage you measure on pin '99a' of an ECU compared to the figure in manufacturer's data – for example. This is the right way to work as it is an important part of the diagnostic process. If the voltage on pin '99a' reads 12.4 V and data says 12.3 to 12.7 V, then you can be confident that you have a correct reading and take the next step.

However, what do you do if the reading you take is 12.2 V? Or what if the waveform is just slightly different from the known good one?

Sorry, no perfect answer to this one! You will need to use your experience. One key thing, though, is to consider how accurate your readings are and how much effect a small variation would have. In the example given, 12.2 V would be okay with me – but I would keep it in the back of my mind, just in case.

9.3.3 **COP and signal probe**

The coil-on-plug (COP) and signal probe from Pico Technology is the easiest and fastest nonintrusive way to check COP ignition coils and spark plugs. However, it is also now very useful for EV systems.

For example, to check the charger operation, it is only necessary to hold the probe near the charging plug (Figure 9.40) and a 50Hz signal

Figure 9.40 COP probe on the charging plug

Figure 9.41 COP probe on the on-board charger input

Figure 9.43 shows the current being consumed by the positive temperature coefficient (PTC) heater on a vehicle. In this example 15.7 amps. As I varied the temperature settings, the current changed accordingly – a quick and easy way to test this circuit.

An interesting aspect of this test is to calculate the power consumed. Power equals voltage multiplied by current:

$$P = V \times I = 380 \times 15.7 = 5966W \ or \ about \ 6kW$$

Electric vehicle batteries are rated in kilowatt-hours. For example, the battery on my PHEV is rated as 8.7 kWh but with a useable value of about 7.8 kWh (Lithium-ion batteries are never

will be shown (Figure 9.42). On this trace you can also see when the charger switched off. Tests can also be done on other parts of the circuit, but if the cable is very well shielded, the reading is much smaller.

One thing to watch out for is that the probe is very sensitive.

9.3.4 **Current flow: HVAC**

Measuring the current flowing from the high-voltage battery is simple with an amp clamp. This can be done using a meter with a built-in clamp (Figure 9.43), a meter with a separate clamp, or an oscilloscope as outlined later.

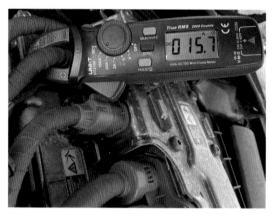

Figure 9.43 Clamp-type ammeter reading the current taken by the car heater

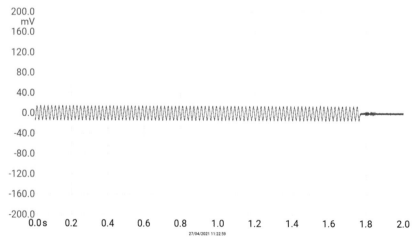

Figure 9.42 A 50Hz waveform from the charging plug via a COP and signal probe

fully charged or fully discharged in order to extend their usable life). The kilowatt-hour is a composite unit of energy equal to one kilowatt (kW) of power sustained for one hour.

Going back to the heater consumption. If this was running for one hour, then it would use 6 kWh – which is a significant proportion of the available 7.8 kWh. It is unlikely that I would need the heater on full power constantly, but it does show why using the heater on an EV causes the available range to drop.

9.3.5 **Current flow: motor**

Measuring the DC current flowing from the high-voltage battery as the vehicle is being driven is a little more difficult than the simple amp clamp meter but not much! For this non-invasive measurement, I will measure the current consumption while driving and the regeneration while braking (decelerating).

I connected an amp clamp around the main high-voltage supply cable as shown in Figure 9.44. Then with some cunning routing of its wire from under the bonnet, through the

Figure 9.44 One of the three amp clamps supplied in the PicoScope EV kit connected around the high-voltage battery supply cable at the inverter

passenger window to the PicoScope, I could set my laptop up on the passenger seat.

The PicoScope vertical scale was set to 40 A/div and the horizontal timebase was set to 10 s/div. A short journey of about 90 seconds, which included normal driving, hard acceleration and braking is shown as Figure 9.45. Where the trace is above the

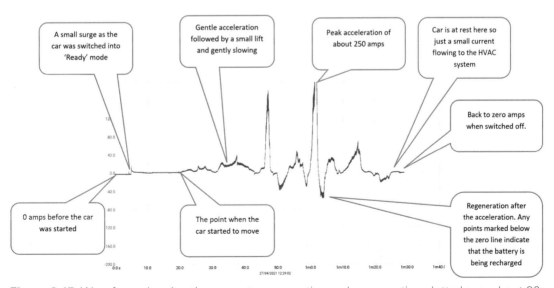

Figure 9.45 Waveform showing the current consumption and regeneration plotted over about 90 seconds while driving

219

zero line, current is being consumed, when the trace is below the zero line, the battery is being recharged.

The peak current was about 250 A and the average current while driving normally on a housing estate road was about 20 A. Let's do another quick calculation here to illustrate another important point:

$$P = V \times I = 380 \times 250 = 95,000W \, or \, about \, 95kW$$

If this was sustained for just 3 minutes (95 x 0.05 hours) it would consume 4.75 kWh, which is more than 60% of the battery capacity on this car. Further, under these extreme conditions, the efficiency drops considerably, too, further compounding the problem. This shows how consumption and, therefore, range is affected by driving style. It has the same effect on an ICE car but is less noticeable because of the much larger fuel quantity and range.

9.3.6 Summary

Contactless and, therefore, non-invasive measurements have a number of advantages on an EV. These are

▶ safety – no high-voltage is exposed
▶ accuracy – the circuit is not affected
▶ speed – quick and easy to connect

Figure 9.46 PicoScope 4425A (visit www. picoauto.com for more information)

Use the best and most appropriate test equipment, and as you become familiar with what readings you expect to get on different EV circuits and you follow a logical process, then diagnostics becomes much easier.

9.4 Charging testing

9.4.1 Protocol

The charging system of an EV is built around standards set by the SAE, known as J1772. The J1772 connector (called the Mennekes in Europe) is designed for single-phase electrical systems with 120 V or 240 V but, with the addition of two extra pins, can also be used with three-phase. The car is connected to an electrical vehicle supply equipment (EVSE) unit that sits between the mains supply and the vehicle. This unit is in communication with the vehicle (when plugged in), which is also defined by the standard.

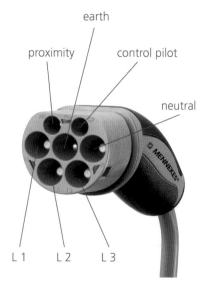

Figure 9.47 Connector has seven pins if three-phase and five if only single phase: Proximity (PP), Protective earth (PE), Control pilot (CP), Line 1 single phase (L1) and neutral (N), Lines L1, L2, L3 and neutral are used for three phase (Source: Mennekes)

9.4.2 Communication

The signalling protocol has been designed so that the EVSE and vehicle can communicate; for example, the charge rate required is determined by the vehicle – up to the limit of what the charger can supply. As the cable is plugged in, the following process is initiated (control pilot functions):

▶ Supply equipment detects vehicle is plugged in.
▶ Supply equipment indicates to vehicle readiness to supply energy.
▶ Vehicle ventilation requirements are determined.
▶ Supply equipment current capacity is provided to vehicle.
▶ Vehicle commands energy flow.
▶ Vehicle and supply equipment continuously monitor continuity of safety ground.
▶ Charge continues as determined by vehicle.
▶ Charge may be interrupted by disconnecting the plug from the vehicle.

At the start of this process, the EVSE puts 12 V on the control pilot (CP) line. On the vehicle, the CP line is permanently connected to the protective earth (PE) via a 2.74 kΩ resistor. This makes the voltage drop to 9 V when a cable is connected, which also starts a 1 kHz pulse-width modulation (PWM) wave generator. Charging is activated by the vehicle adding a 1.3 kΩ resistor in parallel. This results in a voltage drop to 6 V; alternatively, a 270 Ω resistor is used for a required ventilation, resulting in a voltage drop to 3 V. The charging station monitors the voltage on the CP-PE loop. A diode is used in the circuit to make voltage drop in the positive range. A negative voltage indicates a fatal error (like touching the pins), and the power will be switched off.

The status settings are shown as Table 9.2:

EVSE and vehicle handshake sequences are not always the same. However, there is always a change of state between state B, where the

Table 9.2 Status settings

Base status	Voltage, CP-PE (V+/-1)	Charging status
A	0 (car) 12 (EVSE)	Not connected
B	9	Vehicle connected
C	6	Ready for charging
D	3	With ventilation
E	0	No power (shut down)
F	-12	Error

PWM signal peak is at 9 V, and state C, when it drops to 6 V and the vehicle initiates charging.

The PWM duty cycle of the 1 kHz CP signal indicates the maximum allowed mains current. Type 2 devices can also have signals such as LIN or single wire CAN superimposed on the CP line at around 5% duty cycle, and this can set the charge rate. This may make the signal waveform look different and the current may not be as expected. PWM duty cycles are shown in Table 9.3. On some vehicles, all the decisions and settings of charge rate are determined by the on-board charger.

9.4.3 Breakout box

To make connections to the charging circuit between the car and the charger, it was necessary to build a breakout box. I am sure these will become available soon if not already. The one shown is a Type 2 for single phase only.

Table 9.3 Duty cycles

PWM duty cycle (%)	Current (A)
10	6
20	12
30	18
40	24
50	30
66	40
80	48
90	65
94	75
96	80

Figure 9.48 Homemade breakout box (green PE, yellow CP, black N, red L1 vehicle, red L1 EVSE to allow ammeter connection)

9.4.4 **Charging waveforms**

Using the breakout box (Figure 9.49) I connected the EV Automotive PicoScope to the CP line, the mains voltage supply (L1) from the EVSE and also the current in the high-voltage battery cable using an amp clamp.

The blue trace shows the current flowing to charge the battery. In this case about 9 A. The green trace shows the mains supply 50 Hz AC

Figure 9.49 EV PicoScope connected to the breakout box CP line and mains AC supply and via an amp clamp on the main high-voltage battery supply cable. High-voltage protection gloves were used when making the live connections.

at a voltage of 244.6 V, which was connected via a x200 differential probe to isolate high-voltage from the PicoScope. It was a sunny day, and my solar PV panels are the reason the voltage is higher than the normal 230 V mains. The red trace shows the CP line 1 kHz signal at a maximum voltage of 6 V, indicating that charging is taking place.

9.5 **Three-phase testing**

9.5.1 **Operation**

It is acceptable to describe the electrical supply from the inverter to the motor as a three-phase AC sinewave. Likewise, the regenerative current from the motor (during deceleration/braking) can also be described in a similar way. This is because, for most purposes, this is fine and helps us to understand the basic operating principles of an electric drive (simplified):

1. Driver demands power by pressing the accelerator the pedal.
2. Battery high-voltage is inverted into three-phase AC to drive the motor.
3. Driver demands braking by pressing the brake pedal.
4. Car momentum drives the motor as a generator, which produces three-phase AC.
5. This is rectified into DC and used to charge the battery.

In reality, the waveforms are slightly different due to the way they are created. In Figure 9.51 you can just see the small step changes and a ripple. This waveform was captured using three current clamps on the three phase wires connected from the inverter to the motor.

9.5.2 **Motor drive waveforms**

Figures 9.52, 9.53 and 9.54 show the waveforms captured as the car was being driven. The first image shows acceleration from zero, then a steady speed, then accelerating again to a higher speed. The second image shows gentle deceleration (coasting down) and at the end the drop is where I touched the brakes and regeneration

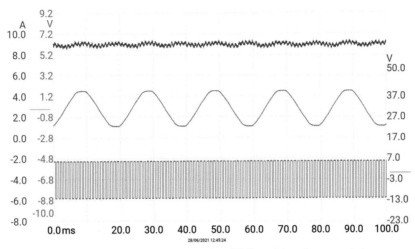

Figure 9.50 PicoScope waveforms captured on the CP line, L1 voltage and the current flowing to the battery

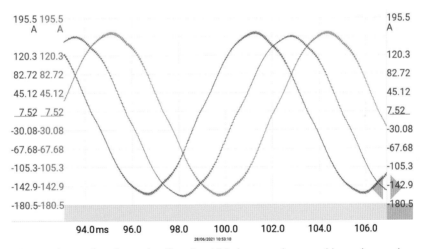

Figure 9.51 Three phase signals each offset by 120 degrees (zoomed in and superimposed)

was initiated. Regeneration reverses the current flow. It doesn't cause the drop in voltage, which in this case was simply because the required braking was small. The third image shows a gradual decrease almost to zero speed during which time regeneration continued. The actual brakes were not activated. The frequency of the signal decreased as the vehicle came to a stop.

9.6 Batteries

9.6.1 Battery repairs

EV battery packs can be repaired in two main ways, depending on the type of repair needed and the specific circumstances:

1. Cell replacement: EV battery packs consist of multiple individual cells, and if a particular cell becomes faulty or degraded, it can

223

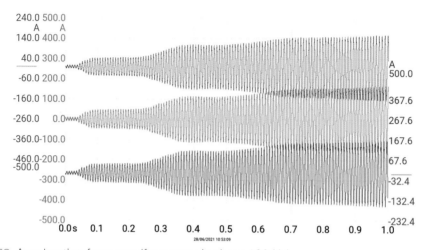

Figure 9.52 Accelerating from rest (frequency is about 120 Hz)

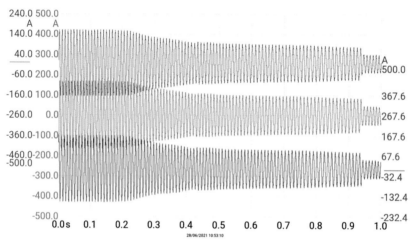

Figure 9.53 Decelerating gently and then regenerating at the end

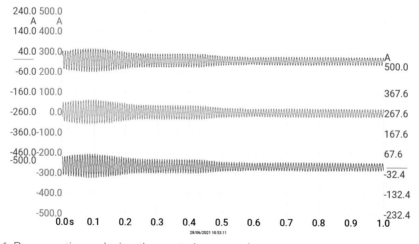

Figure 9.54 Regeneration reducing the car to low speed

be replaced. In this repair method, the damaged cell is identified and removed from the battery pack, and a new cell is installed in its place. This helps restore the battery's overall performance.

2. Module replacement: In some cases, rather than replacing individual cells, an entire module of cells may need to be replaced.

Both of these methods require the faulty cells or packs to be identified, first by scanning and reading the data on voltage, SOC and SOH of the cells (or modules). Once replacements have been determined, the new ones must be balanced to match the existing cells. This requires specialist equipment.

In some cases, the battery's performance is improved by recalibrating or restoring its cells. This procedure involves charging and discharging the battery under controlled conditions to balance the cell voltages and reduce any memory effect or capacity loss. Reconditioning can sometimes help rejuvenate a degraded battery and improve its overall performance.

EV manufacturers often release software updates for their vehicles, including updates specifically for the battery management system (BMS). These updates can optimise the battery's performance, improve charging efficiency and address any known issues or limitations.

In cases where the battery is severely damaged or degraded, it may need to be replaced entirely.

Figure 9.55 shows a high-voltage EV/HEV battery service tool, the Midtronics GRX-5100 performs full-battery pack service for hybrid and electric vehicles:

▶ safely discharges energy from the battery pack
▶ individually balances battery pack modules to ensure optimal pack operation
▶ continuously monitors voltage and temperature during service
▶ on-screen verification that the battery is properly connected and operations can safely begin

Figure 9.55 EV/HEV battery service tool (Source: www.midtronics.com)

▶ captures vehicle VIN and generates a detailed battery pack status report

This equipment is used without removing the pack. A similar kit is available for balancing individual cells or modules after removal from the battery pack.

9.6.2 **Balancing**

Battery cell balancing refers to the process of equalising the state of charge (SOC) or voltage levels of individual cells within a battery pack. In a battery pack, there are multiple cells connected in series or parallel to achieve the desired voltage and capacity. Over time, due to variations in cell characteristics and usage patterns, the cells may experience imbalances in their SOC or voltage levels.

Cell balancing is essential to ensure optimal performance, longevity and safety of the battery pack. Imbalances can lead to several issues, such as reduced overall capacity, decreased energy efficiency, accelerated aging of cells and potential safety hazards.

The cell-balancing process involves redistributing energy or charge between cells to bring them to a similar SOC or voltage level. This can be done through various techniques depending on the battery chemistry and the balancing system used. Common methods include

1. Passive balancing dissipates energy as heat from cells with higher SOC or voltage levels. This is achieved through passive resistors connected in parallel with each cell.

2. Active balancing involves transferring energy between cells by using additional circuitry, such as switching devices or integrated circuits. Energy is transferred from cells with higher SOC or voltage to those with lower levels.

3. Top balancing involves equalising the higher voltage cells in a battery pack, while bottom balancing focuses on equalising the lower voltage cells. These methods are used in certain battery chemistries to optimise the cell-balancing process.

The balancing process is typically controlled by a battery management system (BMS) that monitors and manages the individual cells' SOC, voltage and temperature. The BMS ensures that the cells operate within safe limits and initiates the cell-balancing process, when necessary, either during charging or discharging.

By maintaining cell balance, battery cell balancing helps maximise the overall capacity, extend the battery pack's lifespan, improve energy efficiency and enhance the safety and performance of the battery system as a whole.

9.7 Summary

All the waveforms shown previously are from a correctly operating car so are 'known good'. This is a great way to learn more about how systems work, and, of course, they are a useful comparison that you can use when testing a vehicle with a fault.

A key takeaway note is just how much testing can be done without disconnecting or dismantling components on the vehicle.

EVs are the future, like it or not, so this sort of work will become more common. It is hard to find the time in a busy workshop, but it is a fascinating area of technology – so plug in your scopes and start measuring!

Note

1 The electricity at work regulations 1989, *Regulation* 16, p. 42.

CHAPTER 10

High-voltage pathway for the independent workshop

10.1 Introduction

The automotive industry is witnessing a paradigm shift, transitioning from the once-dominant internal combustion engine (ICE) vehicles to the technologically advanced realm of electric vehicles (EVs) and hybrids. These cutting-edge innovations, coupled with increasing environmental awareness and supportive government policies, have fuelled an explosive growth in the EV and hybrid markets globally. As we step into a new era of sustainable mobility, the role and responsibilities of traditional garage owners are evolving rapidly, necessitating an agile response to stay relevant and profitable.

For many, the mention of EVs and hybrids might evoke images of high-tech gadgets, complex electrical systems, and software-driven functionalities. While it's true that these vehicles are significantly more technologically advanced than their ICE counterparts, at their core, they are still vehicles. They still need to stop, steer and stick to the road – the foundational elements of any vehicle, regardless of the technology under the bonnet.

Figure 10.1 Don't be daunted by change – there is help out there!

In other words, the principles of vehicle maintenance that garage owners have honed over years of working with ICE vehicles still apply. The brakes that allow vehicles to stop safely, the steering systems that ensure precise control and the tyres and suspension that keep the vehicle firmly planted to the road – all these systems remain crucial in EVs and hybrids. Moreover, due to the increased weight and different torque delivery of EVs and hybrids, services such as wheel alignment

DOI: 10.1201/9781003431732-10

and tyre replacement become even more important.

However, there's no denying that EVs and hybrids bring with them a unique set of maintenance needs. Components like batteries, motor controllers, charging systems and regenerative braking systems necessitate a new set of skills and tools. Additionally, the increased reliance on software for controlling various vehicle functions means that issues might not always be mechanical in nature but could also be related to the software. As such, the traditional spanner and screwdriver need to be supplemented with diagnostic tools and software knowledge.

This transition might seem daunting, but it is also a source of immense opportunity. The EV and hybrid market is expanding at an unprecedented rate and shows no signs of slowing down. As per recent studies, EV sales are projected to reach 26 million units annually by 2030, accounting for more than a quarter of the world's car fleet. Therefore, garage owners who adapt their businesses to cater to this emerging market stand to gain a significant competitive advantage.

In this chapter, we aim to guide independent garage owners through this transitional journey. From understanding the EV/hybrid market and adapting business models, to setting up the workspace for EV/hybrid maintenance and building relationships with manufacturers and suppliers, we'll delve into the key steps required to succeed in this new automotive landscape. Let's embark on this exciting journey together, evolving with the industry to ensure that the vehicles of the future continue to stop, steer and stick to the road.

10.2 Analysing the EV/hybrid market

Understanding the market you're serving is the foundation of any business strategy, and transitioning to the maintenance of EVs and

Figure 10.2 An independent workshop can see a wide variety of vehicles with new technologies sitting alongside much older vehicles

hybrids is no exception. This section explores the demographics of EV and hybrid owners, common issues that require garage services and the projected growth of the EV/hybrid market.

10.2.1 Demographics of EV/hybrid owners

The profile of EV and hybrid owners has evolved significantly over the last decade. Early adopters were generally environmentally conscious individuals, driven by the aspiration to reduce their carbon footprint. They were typically well-off, able to afford the higher upfront costs of these vehicles.

Today, the demographics have expanded dramatically. The declining costs of EVs and hybrids, coupled with increasing awareness about climate change, have made these vehicles attractive to a broader range of consumers. Now, the typical EV/hybrid owner could be anyone from a college student wanting a low running cost vehicle, a middle-aged professional concerned about the environment, or a retiree attracted by the lower maintenance needs of these vehicles.

Moreover, as the EV charging infrastructure improves, these vehicles are becoming

Figure 10.3 More local governments operate within ultra-low emissions zones and require vehicles to suit

increasingly appealing to people living in apartments and urban areas who previously may not have considered an EV due to charging constraints. Businesses and government agencies are also jumping on the bandwagon, replacing their fleets with EVs and hybrids to meet sustainability goals and lower operational costs.

10.2.2 Common issues that require garage services

While EVs and hybrids are often lauded for their lower maintenance requirements compared to ICE vehicles, they are not without their unique set of issues and are highly likely to be far more maintenance centric than their ICE forebearers. Understanding these common problems

will help garage owners prepare to offer the right services. Brake fluid changes are more complex for an EV, and quite often specialist equipment to change it is required.

One of the main issues is related to their increased weight and different torque delivery. This puts more strain on the tyres, leading to faster wear and tear. Consequently, regular wheel alignment and tyre rotation and replacement become crucial services for EV and hybrid owners. Misalignment can not only cause uneven tyre wear but also impact the vehicle's range – a key concern for EV owners.

Air-conditioning servicing is another critical service. Unlike ICE vehicles, where the AC is often considered a comfort feature, in many EVs and hybrids, it plays a vital role in cooling the battery and ensuring optimal performance.

Figure 10.4 Brake fluid change machine

Regular servicing is crucial to maintain this system's efficiency and longevity.

Remember

There isn't a belt to drive an air-conditioning compressor for an EV, so these units are often high-voltage and require suitably qualified technicians to work with them.

Moreover, battery health is a major concern for EV and hybrid owners. Over time, the battery's capacity can degrade, affecting the vehicle's range. Garage services will need to offer battery health checks and cell balancing. In some cases, battery replacements could be required.

10.2.3 Projected growth of the EV/hybrid market

The EV and hybrid market has been on a steady growth trajectory, and current projections suggest that this trend is set to continue. This growth is expected to be fuelled by factors such as falling battery costs, improving charging infrastructure, increasing environmental awareness and supportive government policies.

What does this mean for garage owners? Simply put, the demand for EV and hybrid maintenance and repair services will continue to grow exponentially. Traditional garage owners who can adapt their businesses to cater to this market stand to benefit immensely from this trend. Those who can establish themselves as reliable service providers for EVs and hybrids early on will have a significant competitive advantage as the market expands.

10.3 Adapting your business model

With a clear understanding of the EV and hybrid market landscape, the next logical step is to adapt your business model to this new reality. This involves adjusting service prices, offering EV-specific services and fostering relationships with EV and hybrid owners.

10.3.1 Adjusting service prices

In the initial stages of transition, you may find that servicing EVs and hybrids could require more time and specialist knowledge compared to ICE vehicles. This is mainly due to the complexity of their systems, the need for special tools and diagnostic equipment and the safety considerations when working with high-voltage components.

In addition, it's important to recognise that the maintenance cycle for EVs and hybrids is different. For example, they typically require fewer lubricant and other fluid changes (or close to very few in the case of pure EVs) but might need more frequent wheel alignments and tyre rotations and replacements due to their weight and torque characteristics.

Figure 10.5 A drain plug from a failed single-speed box on a BMW i3, mileage 30,000 miles

Consequently, you'll need to adjust your service prices to reflect these realities. You may need to charge more for certain services that require specialist knowledge or tools, while other services may be cheaper due to less frequent need. Additionally, bundling certain services, like wheel alignment and tyre checks, could be a great way to provide value to your customers while ensuring regular business.

10.3.2 Offering EV-specific services

As mentioned earlier, EVs and hybrids come with their own set of maintenance needs. It's crucial to develop a comprehensive understanding of these requirements and offer EV-specific services. Quite often there is a misconception from motorists that these vehicles need little or even no maintenance and repair! This might be a different

experience when explaining about component failure or basic maintenance requirements if this belief is in place.

Wheel alignment and tyre services are key. With their heavier weight and different torque delivery, EVs and hybrids tend to wear out their tyres faster than ICE vehicles. Regular wheel alignment checks and adjustments can help ensure optimal tyre life, improve vehicle handling and even enhance range by reducing rolling resistance.

Air-conditioning services also become critical in the context of EVs and hybrids. In many of these vehicles, the AC system plays a dual role – providing comfort for the occupants and, at times, providing the optimum temperature for the batteries. Regular servicing can help ensure the system's efficiency, thereby preserving the battery's health and performance.

Battery health checks are another important service. Over time, the capacity of an EV's or hybrid's battery can degrade, affecting the vehicle's range. Offering regular battery health checks can help owners understand their battery's state of health and plan for potential replacement if necessary.

Lastly, due to the digital nature of these vehicles, services related to software updates, diagnosing and fixing software glitches and resetting digital service records will be essential. Offering these services can set your garage apart as a one-stop-shop for all EV and hybrid maintenance needs.

10.3.3 Building relationships with EV/hybrid owners

Establishing strong relationships with your new customer base is key to ensuring repeat business and referrals. In today's digital age, customer expectations have evolved; they expect transparency, clear communication and a hassle-free service experience.

To meet these expectations, it's crucial to invest in training your team in customer service

Figure 10.6 Joining networks can help promote your business

skills. This includes being able to explain complex technical issues in a language that customers understand, keeping them informed about the progress of their vehicle's service and being honest and upfront about costs.

Leveraging technology can also help improve the customer experience. For example, using a digital platform to manage bookings, provide service updates and maintain digital service records can greatly enhance convenience for your customers.

Moreover, remember that many EV and hybrid owners are passionate about sustainability. By showcasing your commitment to eco-friendly practices,

Figure 10.7 Digital tools are just as important as mechanical tools in the modern workshop

whether it's through responsible waste disposal, using renewable energy or even just reducing paper usage, you can build a deeper connection with this customer base.

10.4 Setting up the workspace for EV/hybrid maintenance

Successfully transitioning to EV and hybrid maintenance is not just about updating your business model and services, but it also involves modifying your workspace to accommodate the specific needs of these vehicles. This includes understanding high-voltage safety considerations, placing EV-specific equipment optimally and ensuring adequate ventilation for battery work.

10.4.1 High-voltage safety considerations

Working with EVs and hybrids involves dealing with high-voltage systems that, if not handled correctly, can pose serious safety risks. Therefore, it's vital to implement stringent safety procedures and equip your garage with the necessary safety equipment.

All staff should be trained in high-voltage safety procedures, including proper disconnection and reconnection of the high-voltage system before and after service. Emergency procedures should be in place and communicated clearly to all staff members. It's also crucial to ensure that high-voltage components are clearly marked to avoid accidental contact.

Invest in high-voltage personal protective equipment (PPE) such as insulated gloves, safety boots and face shields. Remember that regular rubber gloves or work boots don't offer adequate protection against high-voltage electrical shock.

Moreover, consider acquiring insulated tools that are specifically designed for high-voltage work. These tools are built to withstand high-voltage without conducting electricity, thus protecting the user from electrical shock.

10.4.2 EV-specific equipment placement

Incorporating EV-specific equipment into your workspace requires thoughtful planning to ensure efficient workflow and safety. This equipment can range from specialised diagnostic tools to battery lifters and charging stations. These are covered earlier in the book.

Try to position these tools and equipment in locations that optimise their usage. For instance, battery lifting equipment should be located near where it will be used. The charging station, on the other hand, should be conveniently accessible for charging vehicles or during specialist diagnostic work.

Safety First

There are very few circumstances when a vehicle needs to be charging and work carried out. You must be qualified to do this.

Ensure that high-voltage equipment is stored and used in areas where there is minimal risk of contact with water to avoid the risk of electrical shock. Similarly, keep flammable materials away from battery charging and testing areas.

In terms of workspace layout, strive for a configuration that allows for efficient movement between different tasks, minimises congestion and promotes safety. Remember that some EV-specific tasks may take different timescales to their ICE counterparts, so plan for a workflow that allows other tasks to proceed simultaneously.

10.4.3 Adequate ventilation for battery work

When working with EV batteries, adequate ventilation is paramount. This is because when a battery is damaged, it can emit gases that can be harmful if inhaled or even cause an explosion in confined spaces. As the car is potentially booked in for repair or the history may be unknown, all reasonable care must be taken.

Ensure your workspace has a well-designed ventilation system that can effectively remove these gases and provide fresh air. This is particularly important if you're working in a closed environment or if your workspace is small. Depending on the specifics of your garage, solutions could range from natural ventilation, mechanical extraction systems or a combination of both. This is an area that may also be impacted by legislation, so it is well worth keeping up to date through the government website about ventilation.

Figure 10.8 Whilst reconfiguring the workshop, dollies can serve as a short-term solution, remember to check for correct lifting points!

Figure 10.9 Battery pack work carried out in a well-ventilated workshop

233

During the installation of the ventilation system, consider factors such as the volume of air that needs to be replaced per hour, the size and location of inlet and outlet vents and the direction of airflow. The system should be regularly maintained to ensure it remains effective.

Remember, while it's crucial to ensure adequate ventilation, it's equally important to equip your staff with appropriate personal protective equipment (PPE), such as masks and gloves, when working with batteries. They should also be trained on the risks associated with battery gases and the importance of maintaining the ventilation system.

By taking the time to plan your workspace for EV and hybrid maintenance, you'll not only enhance safety but also improve productivity and efficiency, all of which contribute to the success of your transition.

10.5 Building relationships with suppliers, industry associations and technology providers

As a small independent garage transitioning to EV and hybrid maintenance, forming strategic relationships is crucial to your success. These

Figure 10.10 Sharing experiences at networking events can expand your knowledge quickly

relationships should be not only with parts suppliers but also with industry associations, providers of technical information and garage management system providers. In this section, we'll explore the benefits of these relationships and provide guidance on how to establish them.

10.5.1 Benefits of relationships

Parts suppliers: Establishing strong relationships with reliable parts suppliers can grant your garage access to competitive pricing and availability. This can significantly reduce your operation costs and ensure timely services, thus improving customer satisfaction.

Industry associations: Associations like the Institute of the Motor Industry (IMI) play a pivotal role in shaping industry standards and legislation. Being associated with them can provide your garage with valuable insights into industry trends, upcoming legislative changes and qualification and apprenticeship resources. This information can help you stay ahead and ensure your services align with industry best practices.

Technical information providers: Technical information is key to diagnosing and fixing issues in EVs and hybrids. Establishing relationships with providers of technical information can provide your garage with access to up-to-date, comprehensive information that can improve service quality.

Authorised examiner consultants: AECs are very useful for workshops that also offer MOT testing, especially with the emergence of EVs that are smaller than conventional vehicles and may require an examination of classification at first presentation.

Garage management systems: Modern garage management systems are becoming increasingly community-led, fostering a "hive mind" environment where garages can assist each other. Being a part of such a community can provide your garage with valuable insights, advice and assistance.

Figure 10.11 An AEC can be a valuable contributor when understanding EV MOT classifications

10.5.2 How to establish these relationships

Identify potential partners: Start by identifying potential partners who align with your business's needs. Look for reputable parts suppliers who offer competitive prices,

reliable associations like IMI, comprehensive technical information providers and community-led garage management systems.

▶ **Reach out**: Once you've identified potential partners, reach out to them. Express your interest in forming a relationship and discuss how this could be mutually beneficial.

▶ **Negotiate terms**: After initial contact, negotiate the terms of the relationship. This could include pricing, access to information, technical support, for example. Ensure the terms align with your business goals.

▶ **Formalise relationship**: Once the terms are agreed upon, formalise the relationship. This could be in the form of a contract or a mutual agreement.

▶ **Maintain the relationship**: Finally, maintaining the relationship is key. Regular communication, feedback and collaboration can ensure a beneficial long-term relationship.

By forming strategic relationships with suppliers, industry associations and technology

Figure 10.12 Showcase these relationships within signage and on your website

providers, your garage can gain valuable resources, insights and support that can significantly aid your transition to EV and hybrid maintenance.

10.6 Opportunities and challenges

The shift from traditional internal combustion engine (ICE) vehicles to EVs and hybrids brings both opportunities and challenges for independent garage owners. Recognising these will help you strategically position your business for the transition and mitigate potential obstacles.

10.6.1 The growing market for EV repair and maintenance

Transitioning early to servicing EVs and hybrids presents a range of significant opportunities for your garage.

Firstly, the market for EV and hybrid repair and maintenance is rapidly growing. With increasing awareness about environmental sustainability, more consumers are turning to EVs and hybrids. Governments around the world are also pushing this transition, with many setting ambitious targets for phasing out ICE vehicles. This trend is likely to continue,

creating a burgeoning customer base for EV and hybrid services.

Secondly, early adoption of EV and hybrid servicing can give your garage a competitive edge. By being one of the first in your area to offer these services, you can position your garage as a pioneer and a go-to expert for EV and hybrid repairs and maintenance. This can significantly enhance your garage's reputation and customer trust.

Furthermore, transitioning early allows you to build competence over time. EVs and hybrids come with new technologies that require time to understand and master. By starting early, you can learn at a comfortable pace and make mistakes when the stakes are lower.

Finally, offering EV and hybrid services can diversify your revenue streams. These services can be priced premium due to their specialised nature, enhancing your garage's profitability. Although this situation is fast changing, it would be sensible to watch these trends closely.

10.6.2 Potential challenges and how to overcome them

While the opportunities are substantial, the transition to servicing EVs and hybrids also comes with potential challenges.

One of the main concerns is the investment required. Purchasing EV-specific equipment and tools, upgrading your workspace and training your staff can be costly. To mitigate this, consider gradually phasing in these changes to spread out the costs. You could also explore partnerships and affiliations that could offer financial benefits or assistance. Make sure your accounts accurately reflect your investment to maintain tax affairs in order.

Another challenge is the technical complexity of EVs and hybrids. Their advanced electronic systems and high-voltage components can be daunting. However, with comprehensive training and continuous professional development, your staff can become proficient

Figure 10.13 Electric vehicles can come in all shapes and sizes

over time. Relationships with technical information providers can also be invaluable in this regard.

Lastly, there is the challenge of staying up-to-date with the rapidly evolving EV technology. To overcome this, ensure regular training for your staff and keep an eye on industry trends and advancements. Membership in industry associations can also provide access to a wealth of up-to-date information as well as a convenient method of maintaining and recording your CPD.

The transition to EV and hybrid servicing is a significant step, filled with both opportunities and challenges. By recognising these early and proactively addressing them, your garage can smoothly navigate this transition and emerge as a trusted service provider for EVs and hybrids.

10.7 Suggested checklist for transitioning to EV/hybrid maintenance

Transitioning your garage to service EVs and hybrids requires careful planning and execution. This comprehensive checklist outlines the key steps in this process, guiding you from understanding the market to preparing for future growth.

Understanding the market:

▶ Research EV/hybrid market: Study your local and global market for EVs and hybrids. Understand the demographics, buying behaviours and preferences of EV/hybrid owners.

▶ Identify common issues: Familiarise yourself with the common issues that EVs and hybrids encounter. This includes

Figure 10.14 Training event at Delphi in Warwick

understanding the importance of wheel alignment, air conditioning and tyre wear, among others.

▶ Forecast market growth: Research projected growth of the EV/hybrid market. Understand potential customer volume and plan your business capacity accordingly.

Adapting your business model:

▶ Revise service prices: Consider adjusting your service fees to reflect the specialised nature of EV and hybrid services. Ensure your pricing is competitive but profitable.

▶ Offer EV-specific services: Identify and offer services specific to EVs and hybrids, such as battery maintenance and electronic system diagnostics.

▶ Improve customer relations: Develop strategies to build strong relationships with your new customer base. This includes excellent communication, transparency and quality service.

Setting up your workspace:

▶ Prioritise safety: Understand and implement high-voltage safety measures. This includes acquiring safety equipment and training your staff on safe handling of EVs and hybrids.

▶ Arrange workspace: Plan for the integration of EV-specific equipment into your existing workspace. Consider the space and power requirements of these tools.

▶ Ensure adequate ventilation: If you plan on servicing batteries, ensure your garage has adequate ventilation to avoid hazardous fumes.

Figure 10.15 Garage Hive offer a garage management solution but also have a resource for forecasting

Building relationships:

▶ Find reliable suppliers: Establish relationships with reliable parts suppliers for competitive prices and availability.

▶ Join industry associations: Consider joining industry associations like the IMI for insights into legislation, current industry news and qualification and apprenticeship resources.

▶ Partner with information providers: Form relationships with providers of technical information for access to comprehensive, up-to-date information.

▶ Choose a management system: Select a community-led garage management system for community support and collaboration.

▶ Consider attending trade shows: Along with training, trade shows can be a valuable (and enjoyable) time to network and talk to other technicians.

Addressing opportunities and challenges:

▶ Recognise opportunities: Understand the opportunities in the EV/hybrid market and strategise to leverage them.

▶ Overcome challenges: Identify potential challenges in your transition and develop strategies to overcome them. This includes financial planning, training and staying updated on EV technology.

Preparing for future growth:

▶ Plan for expansion: As the EV/hybrid market grows, plan for potential expansion of your business. This could include increasing staff, workspace or service offerings.

▶ Invest in continuous learning: Encourage continuous professional development

Figure 10.16 You bump into all sorts of people on the stands at trade shows!

for your staff to keep up with the rapidly evolving EV technology.

By following a checklist such as this, your garage can smoothly transition to servicing EVs and hybrids, setting itself up for success in the burgeoning EV market.

10.8 Summary

As we reach the conclusion of this chapter, it's vital to reiterate the importance of transitioning your garage services to cater to electric vehicles (EVs) and hybrid electric vehicles (HEVs). As the shift towards sustainable mobility gains momentum worldwide, it's evident that the automotive industry is on the brink of a substantial transformation. This new reality presents an unmissable opportunity for garage owners to adapt and partake in an exhilarating new chapter of automotive history.

Bear in mind the quintessential principle of vehicle maintenance – that every vehicle, irrespective of its power source, needs to 'stop, steer and stick to the road'. Whether it's an EV, an HEV or a traditional ICE vehicle, this tenet holds true. As a garage owner, your proficiency in maintaining these critical systems is a valuable asset that will continue to serve you in this new era of EVs and hybrids.

Indeed, servicing EVs and HEVs involves new challenges like understanding high-voltage systems, electric drive units and advanced battery technologies. However, instead of viewing these as daunting hurdles, see them as opportunities for expanding your expertise and enhancing your service offerings.

Air-conditioning systems, which played a comparatively minor role in ICE vehicles, will take centre stage in EVs and HEVs, given they can have a role in battery thermal management but are often more used to regulate the temperature of the cabin. Regular wheel alignment, critical for energy efficiency, will also gain importance due to the heavier weight of EVs and hybrids. Tyre maintenance and replacement will require renewed attention, considering EVs and hybrids deliver torque differently and carry heavier weight than ICE vehicles.

As the world moves towards digital solutions, handling digital service records smoothly will be another essential aspect of modern garage operations. Similarly, as customer expectations evolve, garage owners must hone their communication skills to match the demands of a well-informed and discerning clientele.

In essence, as the adage goes, 'the business of business has never been so business-like'.

Figure 10.17 An older hybrid car presented for MOT testing

Figure 10.18 Mercedes EV with separate water-cooled systems for the AC system and HV battery pack

This is particularly relevant in the current automotive landscape. The evolution isn't just technological; it encompasses every aspect of your garage operations. As you learn to work on advanced EVs and hybrids, your business practices need to keep pace, ensuring they fulfil modern motorists' expectations.

The switch to EV and hybrid servicing might seem overwhelming, but the early adopters who manage this transition well can position themselves as pioneers in this growing field, earning customer trust and establishing a robust reputation.

Remember, transitioning to EV and hybrid servicing is more a marathon than a sprint. It requires careful planning, investment, continuous learning and, most importantly, adaptation. However, with the right approach and resources, this transition can turn into an exciting evolution, steering your business towards a sustainable and prosperous future.

With the insights shared in this chapter, you are now equipped to embark on this remarkable journey into the world of electric and hybrid vehicles. Embrace the opportunities, adapt your business practices and steer your garage into a promising future, keeping pace with both technology and customer expectations. Your business has never been so business-like, and the road ahead has never been so electrifying.

10.9 Case studies

In this section, we will delve into two real-life case studies of UK garages that have successfully made the transition to servicing EVs and hybrids, offering insights and key lessons.

10.9.1 Lindleys autocentres (Nottingham)

Lindleys Autocentres, a distinguished name in the UK automotive service industry, is an example of an enterprise that recognised

and acted on the rising trend of data-driven decisions. What began as a modest garage in Arnold eventually expanded into a successful chain of branches scattered across Nottingham, thanks to its strategic use of Garage Hive.

Garage Hive, a community-driven garage management system powered by Microsoft Business Intelligence, provided Lindleys Autocentres with an efficient way to manage its operations, from customer bookings to billing. The platform, continuously enhanced by contributions from its user community, including Lindleys, helped Lindleys streamline its operations and leverage data analytics for business insights.

Simultaneously, Lindleys invested in staff training, obtaining the necessary certifications to handle high-voltage systems safely, thereby readying its team for the influx of EV and hybrid vehicles. It also ensured the acquisition of specialised tools and equipment required for EV and hybrid repairs and services. Lindleys upheld its guiding principle of 'stop, steer and stick to the road', applying it to its approach to EVs and hybrids as well.

Lindleys Autocentres' successful transition underscores the importance of embracing modern business tools while upskilling and adapting to new technologies in the automotive industry.

10.9.2 Cleevely EV (Cheltenham)

Cleevely EV in Cheltenham stands as a testimony to successful adaptation and foresight. A part of the family-owned Cleevely Motors, the garage underwent a transformation to become one of the first independent businesses in the UK to specialise in EVs and hybrids. It was the third generation, Matt Cleevely, who saw the potential of the EV market and took the plunge.

Under Matt's leadership, Cleevely EV became an IMI-certified EV service centre, with

241

Figure 10.19 Lindleys has won a number of prestigious awards

Figure 10.20 Proprietor Matt Cleevely (right) directing a diagnostic approach

Matt himself becoming an IMI-certified EV technician. This investment in training and certification endowed the team with the skills and expertise required to service and repair EVs and hybrids.

Cleevely EV also reached out to the community for knowledge sharing and technical support. It fostered relationships with parts suppliers and technical information providers to ensure its services remained current and comprehensive.

The Cleevely EV story teaches us that seizing opportunities, making strategic investments in skills and nurturing relationships can propel a garage to a leadership position in the EV service industry.

These case studies present valuable lessons for any garage planning to venture into the EV and hybrid sector. They demonstrate that the road to transition, while challenging, can lead to success through strategic planning, training, adaptation and efficient business management.

Index

Index